Strategies for Sustainability

Series Editors

Rodrigo Lozano
Faculty of Engineering and Sustainable Development
University of Gävle
Gävle, Sweden

Angela Carpenter
Faculty of Engineering and Sustainable Development
University of Gävle
Gävle, Sweden

The series focuses on "implementation strategies and responses" to sustainability problems – at the organizational, local, national, and global levels.

Our objective is to encourage policy proposals and prescriptive thinking on topics such as: sustainability management, sustainability strategies, lifestyle changes, regional approaches, organisational changes for sustainability, educational approaches, pollution prevention, clean technologies, multilateral treaty-making, sustainability guidelines and standards, sustainability assessment and reporting, the role of scientific analysis in decision-making, implementation of public-private partnerships for resource management, regulatory enforcement, and approaches to meeting inter-generational obligations regarding the management of common resources.

We favour trans-disciplinary perspectives and analyses grounded in careful, comparative studies of practice, demonstrations, or policy reforms. This largely excludes further documentation of problems, and prescriptive pieces that are not grounded in practice, or sustainability studies. Philosophically, we prefer an open-minded pragmatism – "show us what works and why" – rather than a bias toward a theory of the liberal state (i.e. "command-and-control") or a theory of markets. We invite contributions that are innovative, creative, and go beyond the 'business as usual' approaches.

We invite Authors to submit manuscripts that:

- Document and analyse what has and has not worked in practice;
- Develop implementation strategies and examine the effectiveness of specific sustainability strategies;
- Propose what should be tried next to promote greater sustainability in natural resource management, energy production, housing design and development, industrial reorganization, infrastructure planning, land use, business strategy, and organisational changes;
- Prescribe how to do better at incorporating concerns about sustainability into organisations, private action, and public policy;
- Focus on trans-disciplinary analyses grounded in careful, comparative studies of practice or policy reform; and
- Provide an approach "...to meeting the needs of the present without compromising the ability of future generations to meet their own needs," and do this in a way that balances the goal of economic development with due consideration for environmental protection, social progress, and individual rights.

Themes covered in the series are:

Sustainability management
Sustainability strategies
Lifestyle changes
Regional approaches
Organisational changes for sustainability
Educational approaches
Pollution prevention
Clean technologies
Multilateral treaty-making
Sustainability guidelines and standards
Sustainability assessment and reporting
The role of scientific analysis in decision-making
Implementation of public-private partnerships for resource management
Governance and regulatory enforcement
Approaches to meeting inter-generational obligations regarding the management of common resources

More information about this series at https://link.springer.com/bookseries/8584

Rodrigo Lozano

Toward Sustainable Organisations

A Holistic Perspective on Implementation Efforts

Springer

Rodrigo Lozano
Faculty of Engineering and Sustainable Development
University of Gävle
Gävle, Sweden

ISSN 2212-5450 ISSN 2452-1582 (electronic)
Strategies for Sustainability
ISBN 978-3-030-99675-8 ISBN 978-3-030-99676-5 (eBook)
https://doi.org/10.1007/978-3-030-99676-5

This Springer imprint is published by the registered company Springer Nature Switzerland AG
The registered company address is: Gewerbestrasse 11, 6330 Cham, Switzerland

To our future change agents, may this book serve as a stepping stone to great changes.

To Dave: This book would not have been possible without your input. You were a great teacher and mentor. We will miss your great knowledge, experience, elegance, kindness, and wit.

Foreword

Two essential prerequisites for all types of organisations to be successful are the need of orientation and a long-term perspective.

Among all the paradigms composing this fundamental set of principles, "sustainability" holds a pivotal position. In essence, it has become virtually impossible to oppose or criticise any concept or resolution that claims to be based on sustainability. On the other hand, considerable efforts have been made to establish an agreed and elaborate understanding of sustainability as a (in Rodrigo Lozano's words) "broad, complex, controversial, open-ended and challenging notion".

Scientific research digested, evaluated, and displayed the plethora of experiences organisations have made in their attempts to integrate "sustainability" into their structure, culture, and, of course, their business models. What we see as an outcome of this process is sustainability finding its way into the established set of fundamental pillars of economic and societal evolution. Based on this wealth of documentation, on this treasure chest of reliable data and findings, we are now in a position to pose a set of key questions to be considered if the implementation of "sustainability" in an organisation should become a true success story:

- Is there an established common understanding on the various dimensions of "sustainability" as well as "sustainable development"?
- Is the implementation of "sustainability" understood as a process of transformation embedded in a holistic strategy of "change management"?
- Is there a best practice for applying "tools, initiatives, and approaches" in the framework of implementing strategies?
- To what extent have specific types of organisations (representing civil society, public service, or corporations) contributed to the development of a common and robust understanding of "sustainable development"?
- As to the criteria for successful implementation: Is there a common understanding on the relevance of external stimuli, internal drivers and hurdles, and the role of partnerships?

This book makes active use of Rodrigo's unique academic skill and experience to offer not only elements for answers and approaches to each of these key questions—he also manages to elaborate and display the very essence in each of these clusters. Against the background of 30 years of my experience in environmental policy at national, European (I had the honour of leading the EU Presidency Team in 1998, 2006, and 2016 and chaired the Management Board of EU's Chemical Agency ECHA for 4 years) as well as international level, I fully associate myself with Rodrigo's approach and key findings. It was a pleasure for me to dig into his analysis, a true milestone in scientific interpretation and guidance for implementing this key paradigm.

Austrian Ministry of Sustainability and Tourism
Wien, Austria
Thomas Jakl

Acknowledgements

This book is the result of many years of research and teaching sustainability in organisations. I would like to thank those who made this book possible, in particular, Maria Barreiro-Gen and David Cromie (who also proofread the manuscript) for their continuous encouragement, inspiration, and pressure in encouraging me write this book.

I would also like to thank my parents for giving me the mindset and drive to making sustainability a reality. I would also like to thank all the respondents of the surveys, whose responses have been invaluable in the making of this book.

Contents

1 Introduction 1
1.1 Introduction 1

2 Sustainable Development and Sustainability 5
2.1 Introduction 5
2.2 A Timeline of Sustainable Development 7
2.3 Sustainability Perspectives 11
2.4 SD and Sustainability Discussion 13
References 14

3 Organisations and Their Contribution to Sustainability 19
3.1 Introduction 19
3.2 Organisational System Elements 20
3.3 Organisation Stakeholders 23
3.4 Organisational System Interactions 26
3.5 Organisations' Sustainability Efforts 30
References 33

4 Tools, Initiatives, and Approaches Used for Implementing Sustainability 39
4.1 Introduction 39
4.2 Operations and Production-Oriented TIAs 39
4.2.1 Circular Economy 40
4.2.2 Cleaner Production (CP) 41
4.2.3 Design for the Environment (DfE) 42
4.2.4 Factor X 44
4.2.5 Green Chemistry (GC) 45
4.2.6 Industrial Ecology (IE) 46
4.2.7 Life Cycle Assessment (LCA) and Life Cycle Management (LCM) 47
4.3 Management and Strategy-Oriented TIAs 48
4.3.1 Corporate Citizenship (CC) 48

4.3.2 Corporate Social Responsibility (CSR) 48
4.3.3 Corporate Sustainability (CS) 51
4.3.4 Sustainable/Socially Responsible Investment (SRI) 52
4.3.5 (More) Sustainable Business Models 52
4.3.6 The Natural Step (TNS) 54
4.3.7 The Triple Bottom Line (TBL) 54
4.4 Supply Chain-Oriented TIAs 55
4.4.1 Eco-Labelling 55
4.4.2 Sustainable Marketing (SM) 56
4.4.3 Sustainable Livelihoods (SL) 56
4.4.4 Sustainable Supply Chains (SSC) 57
4.5 Assessment- and Reporting-Oriented TIAs 58
4.5.1 Environmental and Social Accounting (ESA) 58
4.5.2 Environmental Management Systems (EMS) 58
4.5.3 Integrated Management Systems 60
4.5.4 Sustainability Reporting (SR) 60
4.6 TIA 'Discussion' 62
References 63

5 Organisational Change Management for Sustainability 75
5.1 Introduction 75
5.2 Change Typologies 76
5.3 Change Drivers and Resistance to Change 78
5.4 Change Incorporation and Institutionalisation 79
5.5 Organisational Change Management for Sustainability 79
5.5.1 Sustainability Change Drivers and Resistance to Change 81
5.5.2 Sustainability Incorporation and Institutionalisation in Organisations 84
5.6 Discussion 85
References 85

6 Civil Society Organisations' Contributions to Sustainability 89
6.1 Introduction 89
6.2 Sustainability Priorities and Impact 90
6.3 Use of TIAs by CSOs 92
6.4 Sustainability Change Management in CSOs 95
6.5 Sustainability Collaboration in CSOs 98
6.6 Discussion and Conclusions 99
References 100

7 Corporations' Contributions to Sustainability 103
7.1 Introduction 103
7.2 Sustainability Priorities and Impact 104
7.3 Use of TIAs by Corporations 106
7.4 Sustainability Change Management in Corporations 109

7.5 Sustainability Collaboration in Corporations . . . 111
7.6 Discussion and Conclusions . . . 112
References . . . 113

8 Public Sector Organisations' Contributions to Sustainability . . . 115
8.1 Introduction . . . 115
8.2 Sustainability Priorities and Impact . . . 116
8.3 Use of TIAs by PSOs . . . 118
8.4 Sustainability Change Management in PSOs . . . 120
8.5 Sustainability Collaboration in PSOs . . . 123
8.6 Discussion and Conclusions . . . 125
References . . . 126

9 Towards More Sustainable Organisations: Discussions and Conclusions . . . 127
9.1 Introduction . . . 127
9.2 Sustainability Priorities and Impacts . . . 129
9.3 Use of TIAs by PSOs . . . 130
9.4 Sustainability Change Management . . . 132
9.5 Sustainability Collaboration . . . 136
9.6 Conclusions . . . 137
References . . . 138

Index . . . 141

Acronyms

CC	Corporate citizenship
CE	Circular economy
ChL	Chemical leasing
CP	Cleaner production
CS	Corporate sustainability
CSOs	Civil society organisations
CSR	Corporate social responsibility
DfE	Design for the environment
EMAS	EU Eco-management and audit scheme
EMS	Environmental management systems
ESA	Environmental and social accounting
FTSE	First tier sustainability equilibria
GC	Green chemistry
GRI	Global Reporting Initiative
HEIs	Higher education institutions
IE	Industrial ecology
ILO	International Labour Organization
IMS	Integrated management systems
ISO	International Organization for Standardization
LCA	Life cycle assessment
LCM	Life cycle management
MDGs	Millennium Development Goals
MSOS	More sustainability-orientated state
MuSIC	Multidimensional sustainability influence change
NGOs	Non-governmental organisations
PSOs	Public sector organisations
SBMs	Sustainable business models
SD	Sustainable development
SDGs	Sustainable development goals
SM	Sustainable marketing
SL	Sustainable livelihoods

SQ	Status quo
SQN	Status quo novo
SR	Sustainability reporting
SRI	Sustainable/socially responsible investment
SSC	Sustainable supply chains
TBL	Triple bottom line
TIAs	Tools, initiatives, and approaches
TNS	The Natural Step
TQM	Total quality management
TTSE	Two-tiered sustainability equilibria

Chapter 1
Introduction

1.1 Introduction

This book aims to provide a complementary perspective on technocentric and managerial approaches, so that organisations can balance their sustainability priorities and align them with their impacts; use the tools, initiatives, and approaches (TIAs) more efficiently; engage with change processes; and tap into the considerable benefits of entering into collaboration with other organisations. The book is divided into nine chapters.

Chapter 2 reviews the sustainable development (SD) and sustainability concepts. It traces their history from since the first definition of sustainability in 1713 to the current state. The chapter traces this history, where a number of sustainability perspectives can be found: (1) central focus, (2) substitutability of natural capital, (3) constituency, (4) trends; and (5) scope. These perspectives show that SD is a broad, complex, controversial, open-ended, and challenging notion that is open to different and, in many cases mutually exclusive, definitions and interpretations. This has created much controversy, especially since such characteristics makes sustainability difficult to implement or be of practical value. This chapter postulates that SD encompasses four dimensions (economic, environmental, social, and time), as well as their interrelations. The chapter also highlights that SD and sustainability have been used interchangeably, but they are inherently different. SD is the process by which we achieve sustainability, and sustainability is an ideal dynamic state.

Chapter 3 provides a theoretical discussion on organisations' contributions to sustainability. Organisations (divided into civil society organisations (CSOs), corporations, and public sector organisations (PSOs)) are an integral part of modern societies and have been instrumental in driving sustainability. In the last decade, there has been an increasing interest in organisational sustainability, where several definitions have been proposed. The most complete one encompasses the four dimensions of sustainability, the organisation system elements, its stakeholders,

R. Lozano, *Toward Sustainable Organisations*, Strategies for Sustainability,
https://doi.org/10.1007/978-3-030-99676-5_1

change processes and their rate of change, and how inputs are transformed into outputs throughout the supply chain.

Chapter 4 offers a discussion on the main TIAs used by organisations. From the 1970s, a number of TIAs have been developed to foster and implement fostering sustainability. These have evolved from purely "end-of-pipe" solutions towards whole-system approaches, by changing products, processes, services, and systems, so that waste is minimised and resources used more efficiently and effectively, in almost closed loops. Twenty-four TIAs have been widely used and researched, which can be divided according to their focus on (1) operations and production, (2) management and strategy, (3) supply chain and, (4) assessment and reporting.

Chapter 5 provides a theoretical discussion on organisational change management for sustainability. Organisational change aims to move from the current state to one that is more desirable, ranging from minor to radical changes. Organisational change management is most relevant for organisational sustainability, since it addresses the time dimension by providing a dynamic perspective. However, few organisations have successfully incorporated sustainability into their systems and culture because many sustainability approaches have been based on technocentric solutions and managerial ploys, which neglect culture change processes. When addressing sustainability change, organisations are influenced by their stakeholders. Change management for sustainability must encompass all the system elements in a process where the status quo is disrupted and a new, more sustainability-oriented state is achieved. In this process, the drivers to change must be fostered, and resistance to change must be overcome, in order to institutionalise the changes. The process is iterative, multilevelled, and highly dynamic.

Chapter 6 focuses on CSOs, which have an important role in society by providing education, spiritual well-being, and other services. CSOs have engaged with sustainability for a number of years, particularly through the efforts of higher education institutions (HEIs) and by some religious organisations. The chapter provides insights from empirical research on how CSOs have been engaging and contributing to sustainability (i.e. priorities and impacts) through TIAs, change management, and collaboration.

Chapter 7 focuses on corporations, which provide products and services in exchange for profit. Corporations have engaged with sustainability for several years. The chapter provides insights from empirical research into how corporations have been engaging and contributing to sustainability (i.e. priorities and impacts) through TIAs, change management, and collaboration.

Chapter 8 focuses on PSOs. PSOs play an important role in society by providing a variety of services while connecting the government to corporations and civil society organisations. PSOs have engaged with sustainability for a number of years. The chapter provides insights from empirical research into how PSOs have been engaging and contributing to sustainability (i.e. priorities and impacts) through tools, initiatives, and approaches (TIAs), change management, and collaboration.

Chapter 9 provides the discussion and conclusions by comparing CSOs', corporations', and PSOs' sustainability priorities and impacts, use of the TIAs, engagement with change processes, and benefits and challenges of collaboration. The chapter highlights that there are a number of similarities between organisation types but some considerable differences. However, regardless of their type, organisations should recognise that embedding and contributing to sustainability has to encompass the four dimensions of sustainability (economic, environmental, social, and time); use the appropriate combination of TIAs depending on the organisation type and context; engage the system elements throughout the change management processes, recognising external stimuli, internal factors, drivers for change, incorporation, and institutionalisation; and foster collaboration.

This book provides a complementary perspective on technocentric and managerial approaches, so that organisations may better balance their sustainability priorities and align them with their impacts, use the TIAs more efficiently, engage with change processes, and tap into the considerable benefits of entering into collaboration with other organisations.

This book also presents the theoretical foundation, and the empirical data, for sustainability priorities and impacts, the TIAs, change management, and collaboration in order to showcase the efforts CSOs, corporations, and PSOs have undertaken to contribute to sustainability.

This book thus provides practical solutions to engaging successfully with sustainability. This is achieved through taking an integrative, holistic perspective on how organisations can contribute to sustainability through analyses of sustainability priorities and impacts, TIAs, change management, and collaboration.

Chapter 2
Sustainable Development and Sustainability

2.1 Introduction

This chapter discusses the concepts of sustainable development (SD) and sustainability. It starts by providing a context of the positive and negative impacts that industrialisation and economic changes have brought about. The second section presents a timeline of the most important SD and sustainability milestones. The third section discusses five different SD and sustainability perspectives. The fourth section provides a discussion of the terms for the rest of the book's chapters.

During the last century, many radical changes and innovations have taken place, "…we have unlocked countless secrets of matter, space, and biology… dominated the planet with our numbers, technology, and sophisticated organisations" (Korten, 2001). In particular during the last five decades, there have been unprecedented advances in development and industrialisation with dramatic economic and technological changes (Dunphy, Griffiths, & Benn, 2003; Jensen, 1993; NGLS & UNRISD, 2002; WCED, 1987; World Bank, 2021).

Such changes have had positive and negative impacts on the world. Some of the positive impacts include, for example, a large increase in economic activity, from $2.4 trillion in 1900 to $6.3 trillion in 1950 to $46 trillion in 2001, a 19-fold increase (Brown et al., 2002; Flavin, 2001), the large growth in the output of industrial production, over 50 times during the past century—four-fifths of this since the 1950s (WCED, 1987). Then there is the rise in individual average incomes, from $2582 in 1950 to $7454 in 2001 (Brown et al., 2002; Roodman, 2002), at the same time as a worldwide reduction in working hours from 60 per week in the late nineteenth century, to 55 h at the start of World War I, to the current 40–48 h per week (Gilmore, 2021). Then there is the growth in the international goods trade, from $311 million in 1950 to $5.5 trillion in 2000 (Miller, 2002), and the improvements in agricultural output, the so-called Green Revolution, with the help of synthetic organophosphate fertilisers, from 14 million tons in 1950 to 134 million in 2000 (Brown et al., 2002).

R. Lozano, *Toward Sustainable Organisations*, Strategies for Sustainability,
https://doi.org/10.1007/978-3-030-99676-5_2

These advances have also helped to increase life expectancy (up by 20 years in developing countries), a halving of infant mortality rates, increased literacy rates, increased food production and consumption at rates higher than population growth rates, improved incomes, and the spread of democratically elected governments (Carley & Christie, 2000; Dalal-Clayton & Bass, 2002; WCED, 1987). This is showcased by GDP per capita in the world going up (in 2011 USD) from 1174 in 1820 to 2212 in 1920 and from 8222 to 14,700 in 2016 (Bolt & van Zanden, 2021). Access to electricity has increased from 78% in 2000 to 90% in 2019, while cereal yield per hectare went from 1238 tonnes in 1961 to 4498 in 2018. The percentage of children (12 to 23 months) inoculated against measles has gone up from 12.74% in 1980 to 85.69% in 2019, while the percentage of children out of school has reduced from 28% in 1971 to 8% in 2019, and poverty (at $1.90 a day (2011 PPP)) went from 42.7% in 1980 to 9.3% in 2017 (World Bank, 2021).

In spite of great advances, economic activities have often resulted in detrimental impacts on, or increased risks, to the economic, environmental, and social dimensions (WCED, 1987). This has raised concerns that the resulting damage to the earth's environment and quality of life for future generations will be irreparable (Ayres, 2004; Cairns Jr., 2004; Carley & Christie, 2000; WCED, 1987), especially if developing countries follow the historical Western development model and repeat the same environmental mistakes (Hart, 2000). The world's population has increased from around 3 billion in 1960 to 7.75 billion in 2020, and the prevalence of overweight has increased from 21.5% in 1975 to 38.9% in 2016, as per capita consumption has gone up from 3.12 in 1960 to 4.48 in 2018 (World Bank, 2021), but 70% of the 1.3 billion people living in conditions of poverty are women (Osman-Elasha, 2021), while epidemics (e.g. HIV/AIDS, and COVID-19) have caused major economic and social disparities (Barreiro-Gen et al., 2020), and the gender gap will be closed in 108 years moving at the current pace.

The impact of human activity on the environment has been greater during the last 100 years than in all the previous centuries (Gremmen & Jacobs, 1997), e.g. carbon dioxide (CO_2) emissions went from 9,463,838 kt in 1960 to 34,041,045 kt in 2018 (almost 3.6 times), as seen in Fig. 2.1, where the largest increase has been in upper middle income countries (Fig. 2.2) (World Bank, 2021). There has also been an increase of biodiversity loss, as shown in Fig. 2.3, with an index of 1 for the world in 1970 and 0.58 (weighted) in 2010 (van Goethem & van Landen, 2021), every year 75 billion tons of fertile soil is lost to land degradation (UN, 1994; UNCCD, 2021), and the abundance of species population has declined 44% (van Goethem & van Landen, 2021).

During the last century, economic growth has exacerbated old trends and challenges to economic, environmental, social, and cross-cutting issues (i.e. connecting the three other dimensions), as presented in Table 2.1.

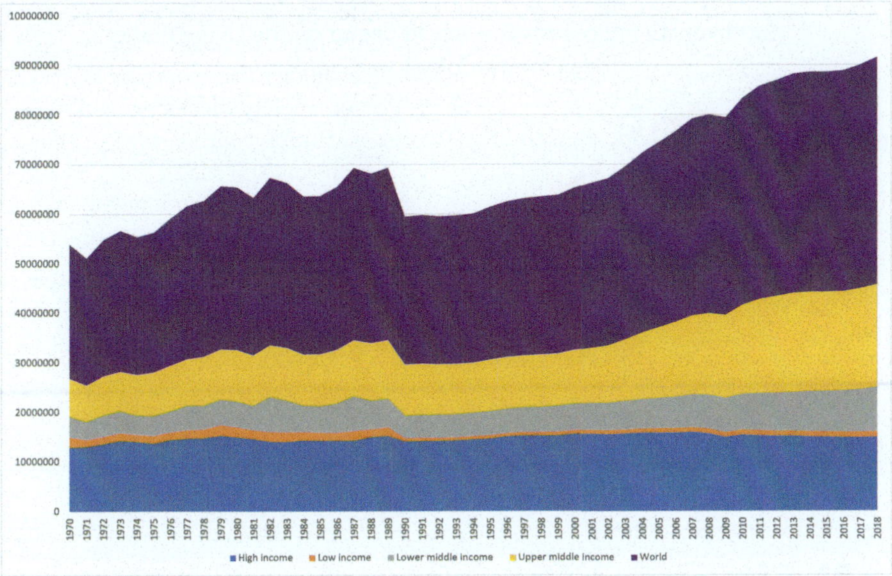

Fig. 2.1 Total greenhouse gas emissions (kt of CO_2 equivalent). Data sourced from climatewatchdata.org/ghg-emissions

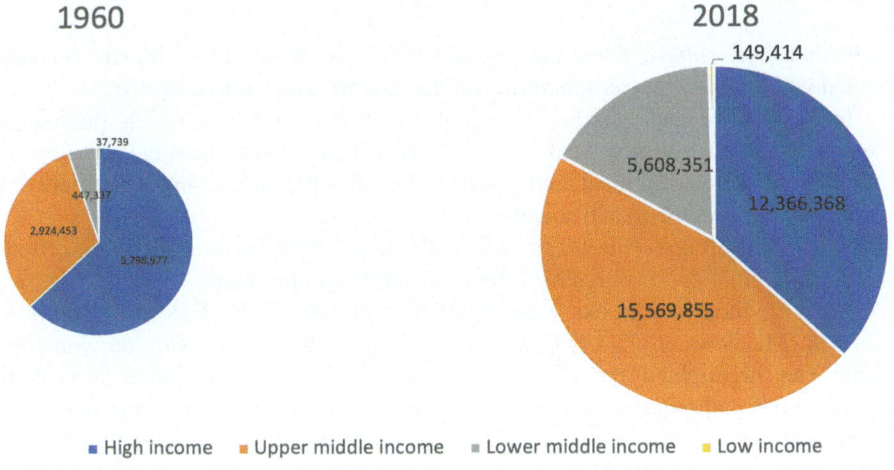

Fig. 2.2 Carbon dioxide emissions (kt) in 1960 and 2018 per economic region. Data sourced from climatewatchdata.org/ghg-emissions

2.2 A Timeline of Sustainable Development

A timeline of the evolution of SD is presented in Fig. 2.4. The timeline shows that SD has its roots in sustainable forest management which was developed in Europe during the seventeenth and eighteenth centuries. In 1713, Hans Carl von Carlowitz

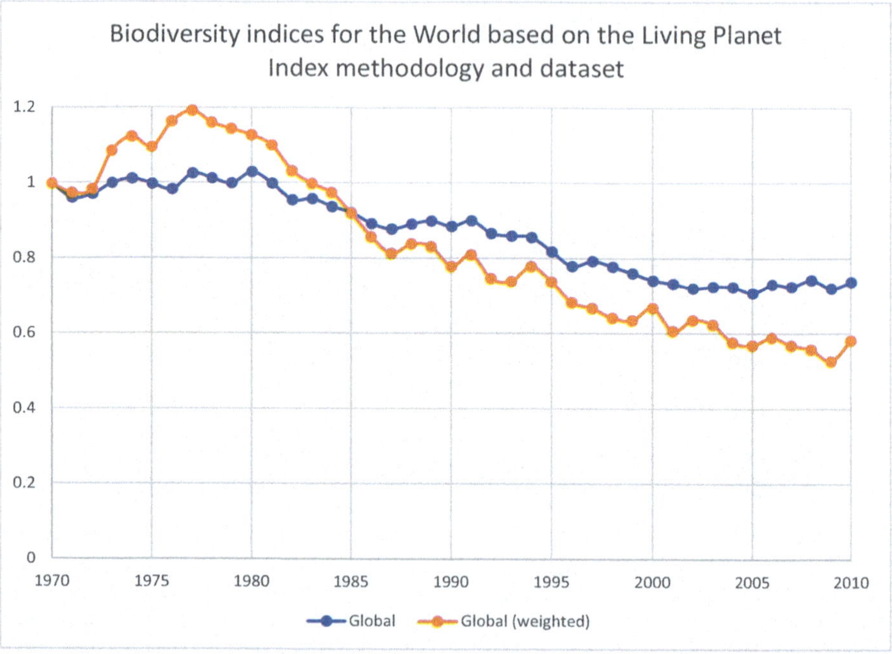

Fig. 2.3 Biodiversity based on the Living Planet Index methodology and dataset

published *Sylvicultura Oeconomica*, which discussed managing forests for sustained yield, under the German term *Nachhaltigkeit* (von Carlowitz, 2013).

In 1962, the book *Silent Spring* was published (Carson, 1962), which sparked the "environmental revolution", exposed the toxic effects of agrochemical products on humans and the environment, and started to make evident interconnections between economic, environment, and social issues.

In 1968, the book *Population Bomb* (Ehrlich, 1968) exposed the connections between population, resource exploitation, and the environment.

In 1972, three key books were published: (1) the Club of Rome's *Limits to Growth* (Meadows et al., 1974) which made clear that resources in the world are finite and highlighted the consequences of continuing exponential growth in resource use and pollution creation; (2) *A Blueprint for Survival* article published (Goldsmith et al., 1972), where the "sustainable" adjective, i.e. capable of being sustained, from the Latin *sustinere*, *sus*- "sub" and *tenere* "hold", was first linked to industrial expansionism and its effects on the environment; and (3) the *UN Conference on Human Environment* (UNEP, 1972) held in Stockholm, which discussed the protection of the environment and its relation to development were for the first time systematically addressed and had a strong focus on regulating development within natural limits set by the carrying capacities of the Planet (Dobers & Wolff, 2000).

Table 2.1 World challenges exacerbated or caused by industrialisation during the last century

Economic dimension issues	Environmental dimension issues	Social dimension issues
1. Economic disparity and political instability 2. Marginalisation 3. Consumption 4. Bribery and corruption 5. Disproportionate income distribution, i.e. rich/poor ratios (within countries, and between developed and developing countries 6. External debt (mainly of developing countries)	7. Global energy use and security 8. Climate change 9. Nitrogen loading 10. Natural resource deterioration 11. Loss of biodiversity 12. Pollution 13. Growing water scarcity 14. Other urban problems 15. Desertification 16. Deforestation and soil degradation 17. Unsafe groundwater 18. Artificial chemicals 19. Global warming	20. Poverty and extreme poverty 21. Undernourishment and food security 22. Diseases and epidemics (e.g. COVID-19, HIV-AIDS, malaria) 23. Population growth 24. Aging population 25. Illiteracy 26. Hunger 27. Gender differences 28. Arms trade and warfare
Cross-cutting issues		
29. Interrelatedness among economic, environmental, and social problems 30. Short-, long-, and longer-term effects and interrelatedness		

Source: Adapted from (Benn et al., 2006; Carley & Christie, 2000; Dalal-Clayton & Bass, 2002; Lozano, 2009; OECD, 2021; UN, 1994; UNCCD, 2021; United Nations, 2018; WCED, 1987; World Bank, 2021)

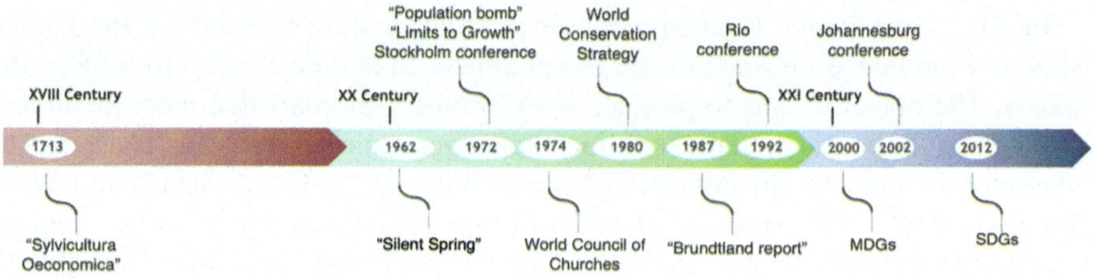

Fig. 2.4 A timeline of the evolution of SD and its key milestones

In 1974, the term "sustainable" was used at the World Council of Churches conference as an adjective in "sustainable society", which was one of the first attempts to explicitly integrate environmental and social issues with economic activities. Accordingly, a sustainable society is defined by four principles: (1) social stability; (2) food supply, carrying capacity and biodegradation; (3) reduction of non-renewable resource use; and (4) human activities that have little or no impact on the global climate.

In 1980, the World Conservation Strategy (IUCN et al., 1980) connected the concept of sustainable society to sustainable development (SD), which was defined as "For development to be sustainable it must take account of social and ecological factors, as well as economic ones; of the living and non-living resource base; and of the long term as well as the short term advantages and disadvantages of alternative actions" (IUCN et al., 1980).

In 1987, *Our Common Future* (WCED, 1987), the Brundtland Report, was published, which proposed a definition of sustainable development: "Humanity has the ability to make development sustainable—to ensure that it meets the needs of the present without compromising the ability of future generations to meet their own needs". This definition has been the most cited (Lozano, 2008; Vos, 2007). The Brundtland Report brought SD to the mainstream international political agenda, thereby raising worldwide awareness (Cheney et al., 2004; Clark, 2001; Stavins et al., 2003).

In 1992, United Nations Conference on Environment and Development (UNCED) (UN, 1997), Earth Summit, was held in Rio de Janeiro, which provided a forum to express global concerns about environmental and developmental issues. The main outcomes were The Earth Charter and Agenda 21 (a global action plan outlining the priorities for SD for the twenty-first century).

In 2000, the Millennium Development Goals (MDGs) started a global effort to tackle extreme poverty rates, to halting the spread of HIV/AIDS and providing universal primary education. The MDGs established measurable, universally agreed objectives for tackling extreme poverty and hunger, preventing deadly diseases, and expanding primary education to all children, among other development priorities with the target date of 2015 (UN, 2005).

In 2002, the World Summit on Sustainable Development took place in Johannesburg (UN, 2002), where it highlighted the urgent world problems of poverty, water, consumption and production patterns, natural resources, and the rich/poor increasing gap. Its main outcome was a plan for SD implementation.

In 2012, Sustainable Development Goals (SDGs) were launched at the United Nations Conference on Sustainable Development in Rio de Janeiro to replace the MDGs. The objective was to produce a set of universal goals that meet the urgent environmental, political, and economic challenges facing our world. There are 17 integrated SDGs: (1) No Poverty; (2) Zero Hunger; (3) Good Health and Well-being; (4) Quality Education; (5) Gender Equality; (6) Clean Water and Sanitation; (7) Affordable and Clean Energy; (8) Decent Work and Economic Growth; (9) Industry, Innovation, and Infrastructure; (10) Reduced Inequality, (11) Sustainable Cities and Communities(12) Responsible Consumption and Production; (13) Climate Action; (14) Life Below Water; (15) Life on Land; (16) Strong Peace and Justice Institutions; and (17) Partnerships to achieve the Goal (United Nations, 2018). It should be remarked that the SDGs were developed by government for government (Meuleman & Niestroy, 2015).

2.3 Sustainability Perspectives

Sustainability discourses have emerged from different perspectives. Some of the most widely recognised are (1) central focus, (2) substitutability of natural capital, (3) constituency, (4) trends, and (5) scope.

It should be noted that some authors depict sustainability as three pillars (economic, environmental, and social); however, "pillars" only connect at the base and roof; therefore, in this book the term "dimension" is used, since dimensions can connect fully to other dimensions.

Within the **central focus** type, it is possible to find the (1) anthropocentric, i.e. considering humans as the centre of the focus (Langer & Schön, 2003); (2) technocentric, which focuses on the transformation of sustainability goals into principles and guidelines, and with the help of technology, they are adopted passively by people into their lifestyles to cope with an imposed built environment (Nguyen Cam, 2004); and (3) eco-centric, where the environment is central and needs to be recovered and protected (Daly, 2002; Mebratu, 1998).

While engaging with the **substitutability of natural capital**, it is possible to find weak and strong sustainability. *Weak sustainability*, the dominant one, takes a more functionalistic approach, utilising negotiation among the different stakeholders to make incremental economic, environmental, and social improvements, while avoiding decreases in total wealth over time (Atkinson, 2000; Bartelmus, 1999; Daly, 2002; Zadek, 1999). Weak sustainability attempts to make the transition smoother through stakeholder negotiation, which might take longer but with fewer conflicts. Negotiations usually take longer than command-and-control approaches (Potoski & Prakash, 2004), with the results seldom visible in the short term.

Strong sustainability tends to be more normative and radical. It proposes greater emphasis on the conservation of natural capital, i.e. keeping it constant, while rejecting the creation of economic value, both social and economic, from its use (Atkinson, 2000; Bartelmus, 1999; Daly, 2002; Zadek, 1999). Strong sustainability prioritises the environment over economic and social aspects, leaning towards the protection of nature. Strong sustainability would be more feasible in an equitable world with low levels of illiteracy, little or no poverty, hunger, corruption and bribery, or gender inequalities.

The third categorisation, **constituency**, offered by Mebratu (1998), is divided into three dimensions:

1. *Institutional*, based on need satisfaction and goal maximisation within three systems: economic, environmental, and social.
2. *Ideological*, based upon philosophies such as modern environmentalism, liberation theology, radical feminism, ecofeminism, eco-socialism, and Marxism. Some of its key concerns are overpopulation and the destruction of natural resources as threats to human survival, the assumption that an ecologically sound capitalist system is a contradiction in terms, human needs can be partly satisfied by non-material interactions with nature, humans behave greedily and corruptly because of the socio-economic system and not because they are inherently like

that, and that nature cannot be dominated but planned and controlled. This dimension's position can bring conflict with stakeholders (e.g. investors, corporations, and certain Western governments) who disagree with Marxist and feministic approaches.

3. *Academic*, based mainly on ecological, sociological, and economic (mainly neoclassical economics, more properly referred to as neoliberal) disciplines (see Dunphy et al., 2003; Guillén, 2001; Rees, 2002). Sustainability is transformed into a commodity in two stages: (1) determining the price of environmental commodities to identify proper levels of environmental protection and (2) transforming the calculated prices into real prices, through environmental taxes, subsidies, or other economic instruments. Some examples include SO_x "cap and trade" and, more recently, CO_2 reduction through the Kyoto Protocol. Two implications are important in this perspective: (1) nature is self-organising, self-reproducing, organic, spatial-temporal, and teleological as a system and (2) humans seek to intervene to obtain their needed resources.

The **trends** category is divided into three major stages: (1) *status quo*, (2) *reform*, and (3) *Transformation* (Hopwood et al., 2005).

In the *status quo* stage the need for change is acknowledged, but the problems faced by the environment and society are not insuperable. Adjustments can be made without major changes. Development is considered to be a synonym of economic growth, and this in turn as part of the solution. This view is the dominant one with corporations and governments.

In the *reform* approach, there is acceptance that there are increasing problems, but there is no consideration that a collapse in ecological or social systems is likely or that fundamental change is necessary. The root of the problem is considered to be a lack of knowledge and information. Considerable changes in policies and lifestyles are recognised; however, they can be achieved over time within the present socio-economic structures. The key to such changes is to persuade governments and international organisations to introduce major reforms.

The *transformation* view considers that there are increasing problems in the environment and society and that these are part of how society and humans interrelate with the environment. These problems are inherent to the economic and power structures of society, because they are not primarily concerned with human well-being or environmental welfare. Major transformation is required to avoid crises and possible future collapse.

The transformation view offers to radically modify current economic systems to make them more aligned with sustainability; however, the approaches proposed tend to fall into the "strong" sustainability type, which may limit their application with certain aspects of society, such as corporations. This could be addressed by corporate efforts that would propose "transformation" but under "weak" sustainability.

The **scope** perspective is divided into five approach types (Lozano, 2008): (1) conventional economics, (2) non-environmental degradation, (3) integrational, (4) intergenerational, and (5) holistic.

The *conventional economists'* approach considers sustainability to be a steady state due to economic viability, i.e. sustained growth and self-sufficiency, and it attempts to simplify into economic terms natural and social phenomena (see Goldin & Winters, 1995; Stavins et al., 2003).

The *non-environmental degradation* approach considers that resources are scarce, increasing consumption cannot be continued indefinitely, natural resources should be used without surpassing their carrying capacities, and environmental capital should not be depleted (see Costanza, 1991; Daly, 2002; Hart, 2000; Miller, 2002).

The *integrational approach* (mostly represented by the triple bottom line; see Sect. 4.3.7), aims to address the three "pillars of sustainability", i.e. the economic, environmental, and social dimensions (as discussed by Cairns Jr., 2004; Dalal-Clayton & Bass, 2002; Diesendorf, 2000; Elkington, 2002; Langer & Schön, 2003).

The *intergenerational approach* is based on considering past, present, and future generations, as well as their interactions, i.e. the *time* dimension, represented by the Brundtland Report definition (as highlighted by WCED, 1987).

The *holistic approach* proposes two dynamic and simultaneous equilibria, the Two-Tiered Sustainability Equilibria (TTSE): the First Tier Sustainability Equilibrium (FTSE) is a depiction of the interactions of three dimensions, the economic, environmental, and social, in the present. The TTSE incorporates the fourth dimension, time, where the FTSE interacts dynamically with the dimensions in the future (i.e. the short-, long-, and longer-term perspectives). For example, the economic dimensions of today interrelate with the economic dimensions of the future but also with the environmental and social dimensions of the present and the future.

2.4 SD and Sustainability Discussion

SD has been considered to be a broad, complex, controversial, open-ended, and challenging notion that is open to different and in many cases mutually exclusive, definitions, and interpretations (e.g. Atkinson, 2000; Gremmen & Jacobs, 1997; Hussey et al., 2001). This has created much controversy, especially since such characteristics make SD difficult to implement and be of practical value (Gremmen & Jacobs, 1997; Hussey et al., 2001; Lozano, 2008), and conflicts from different interpretation in different sectors (Gremmen & Jacobs, 1997).

It has been proposed that the integration of SD is affected by four systems (politics, peace and security, cultural values, and institutional/administrative arrangements) in three contexts (local, national, and global), as shown in Fig. 2.5 (Dalal-Clayton & Bass, 2002). Technology should be added as the fifth system (Nguyen Cam, 2004) and the organisational level to the context level (Lozano, 2018). It should be noted that this perspective does not consider the time dimension, which is essential to the implementation of SD.

In many cases, SD is considered only to address the environmental dimension. In other cases, it focuses on the three "pillars" (economic, environmental, and social). For the purposes of this book, SD encompasses four dimensions (economic,

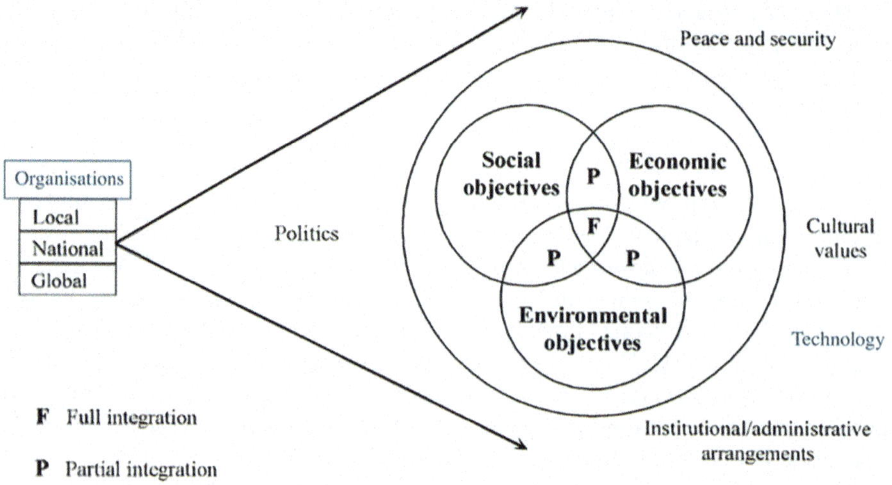

Fig. 2.5 The context and systems of SD (adding organisations to the context and technology to the systems). Source: adapted from (Dalal-Clayton & Bass, 2002)

environmental, social, and time), as well as their interrelations (Lozano, 2008). Following SD is considered to be "A process in which societies continuously strive to improve their quality of life for this generation and future ones, while protecting and improving the natural environment; through changes in activities and attitudes"(Lozano, 2009).

It is common to find the terms "SD" and "sustainability" used interchangeably in the literature, but they are inherently different (e.g. Diesendorf, 2000; Langer & Schön, 2003; Reid, 1995). SD is the process through which sustainability is achieved (Martin, 2003), while sustainability is an ideal dynamic state. The terms are also used contextually; at the institutional and governmental levels, SD is preferred, whereas at the organisational level, sustainability is more widely used.

References

Atkinson, G. (2000). Measuring corporate sustainability. *Journal of Environmental Planning and Management, 43*(2), 235–252.

Ayres, E. (2004). The hidden shame of the global industrial economy. *Worldwatch, 17*, 19–29.

Barreiro-Gen, M., Lozano, R., & Zafar, A. (2020). Changes in sustainability priorities in Organisations due to the COVID-19 outbreak: Averting environmental rebound effects on society. *Sustainability, 12*(12), 5031. https://doi.org/10.3390/su12125031

Bartelmus, P. (1999). *Sustainable development - paradigm or paranoia?* Wuppertal Institute.

Benn, S., Dunphy, D., & Griffiths, A. (2006). Enabling change for corporate sustainability: An integrated perspective. *Australasian Journal of Environmental Management., 13*(3), 156–165. https://doi.org/10.1080/14486563.2006.9725129

Bolt, J., & van Zanden, J. L. (2021). The long view on economic growth: New estimates of GDP. In *How Was Life? Volume II: New Perspectives on Well-being and Global Inequality since 1820*. OECD.

Brown, L. R., Larsen, J., & Fischlowitz-Roberts, B. (2002). *The earth policy reader*. W. W. Norton & Company.

Cairns, J., Jr. (2004). Will the real sustainability concept please stand up? *Ethics in Science and Environmental Politics, 22*, 49–52.

Carley, M., & Christie, I. (2000). *Managing sustainable development* (second). Earthscan Publications Ltd. www.earthscan.co.uk.

Carson, R. (1962). *Silent Spring*. Houghton Mifflin.

Cheney, H., Nheu, N., & Vecellio, L. (2004). Sustainability as social change: Values and power in sustainability discourse. In *Sustainability and Social Science: Round Table Proceedings*.

Clark, W. C. (2001). Research Systems for a Transition toward Sustainability. In W. Steffen, J. Jäger, D. J. Carson, & C. Bradshaw (Eds.), *Challenges of a changing earth*.

Costanza, R. (1991). *Ecological economics. The science and management of sustainability*. Columbia University Press.

Dalal-Clayton, B., & Bass, S. (2002). *Sustainable development strategies: A resource book* (Organization for Economic co-operation and Development (ed.); Paris and). Earthscan Publications Ltd.

Daly, H. E. (2002). Sustainable development: Definitions, principles, policies. In *The future of sustainability*. Springer.

Diesendorf, M. (2000). Sustainability and sustainable development. In D. Dunphy, J. Benveniste, A. Griffiths, & P. Sutton (Eds.), *Sustainability: The corporate challenge of the 21st century* (Vol. 2, pp. 19–37). Allen & Unwin.

Dobers, P., & Wolff, R. (2000). Competing with "soft" issues - from managing the environment to sustainable business strategies. *Business Strategy and the Environment, 9*, 143–150.

Dunphy, D., Griffiths, A., & Benn, S. (2003). *Organizational change for corporate sustainability* (Third). Routledge.

Ehrlich, P. R. (1968). *The population bomb*. Ballantine Books.

Elkington, J. (2002). *Cannibals with forks*. Capstone Publishing Limited.

Flavin, C. (2001). Rich planet, poor planet. In L. Starke (Ed.), *State of the world 2001*. W. W. Norton & Company.

Gilmore, O. (2021). The working week in manufacturing since 1820. In *How was life? Volume II: New perspectives on well-being and global inequality since 1820*. OECD.

Goldin, I., & Winters, L. A. (1995). *The economics of sustainable development*. OECD Cambridge University Press.

Goldsmith, E., Allen, R., Allaby, M., Davoll, J., & Lawrence, S. (1972). A blueprint for survival. *The Ecologist, 2*(1), 1–43.

Gremmen, B., & Jacobs, J. (1997). Understanding sustainability. *Man and World, 30*, 315–327.

Guillén, M. F. (2001). Is globalization civilizing, destructive or feeble? A critique of five key debates in the social science literature. *Annual Review of Sociology, 27*, 235–260.

Hart, S. (2000). Beyond greening: Strategies for a sustainable world. *Harvard Business Review on Business and the Environment, 75*(1), 66–76.

Hopwood, B., Mellor, M., & O'Brien, G. (2005). Sustainable development: mapping different approaches. *Sustainable Development, 13*(1), 38–52. https://doi.org/10.1002/sd.244

Hussey, D. M., Kirsop, P. L., & Meissen, R. E. (2001). Global reporting initiative guidelines: An evaluation of sustainable development metrics for industry. *Environmental Quality Management*, 1–20.

IUCN, UNEP, & WWF. (1980). *Word Conservation Strategy. Living Resource Conservation for Sustainable Development*. International Union for Conservation of Nature and Natural Resources (IUCN)United Nations Environment Programme (UNEP)World Wildlife Fund (WWF).

Jensen, M. C. (1993). The modern industrial revolution, exit, and the failure of internal control systems. *The Journal of Finance, 48*(3), 831–880.

Korten, D. C. (2001). *When corporations rule the world* (2nd ed.). Kumarian Press.

Langer, M. E., & Schön, A. (2003). *Enhancing corporate sustainability. A framework based evaluation tools for sustainable development*. Forschungsschwerpunkt Nachhaltigkeit und Umweltmanagement, Wirtschaftsuniversität Wien.

Lozano, R. (2008). Envisioning sustainability three-dimensionally. *Journal of Cleaner Production, 16*(17), 1838–1846. https://doi.org/10.1016/j.jclepro.2008.02.008

Lozano, R. (2009). Orchestrating Organisational change for corporate sustainability. Strategies to overcome resistance to change and to facilitate institutionalization. In *Business school: Vol. PhD* (Issue August). Cardiff University.

Lozano, R. (2018). Proposing a definition and a framework of Organisational sustainability: A review of efforts and a survey of approaches to change. *Sustainability, 10*(4), 1157. https://doi.org/10.3390/su10041157

Martin, S. (2003). Sustainability, systems thinking and professional practice. *Journal of Geography in Higher Education, 29*(1), 79–89.

Meadows, D. H., Meadows, D. L., Randers, J., & Bherens, W. W., III. (1974). *The limits to growth*. Pan Books Ltd.

Mebratu, D. (1998). Sustainability and sustainable development: Historical and conceptual review. *Environmental Impact Assessment Review, 18*, 493–520.

Meuleman, L., & Niestroy, I. (2015). Common but differentiated governance: A metagovernance approach to make the SDGs work. *Sustainability (Switzerland), 7*(9), 12295–12321. https://doi.org/10.3390/su70912295

Miller, G. T. (2002). *Living in the environment* (12th ed.). Thomson Learning, Inc.

NGLS, & UNRISD. (2002). *Voluntary approaches to corporate responsibility. Readings and a resource guide*. UN Non-Governmental Liaison Service.

Nguyen Cam, C. (2004). A conceptual framework for socio-technocentric approach to sustainable development. *International Journal of Technology Management and Sustainable Development, 3*(1), 59–66.

OECD. (2021). How was life? Volume II. In *How was life? Volume II: Vol. II*. OECD. https://doi.org/10.1787/3d96efc5-en

Osman-Elasha, B. (2021). *Women... in the shadow of climate change*.

Potoski, M., & Prakash, A. (2004). The regulation dilemma: Cooperation and conflict in environmental governance. *Public Administration Review, 64*(2), 153–163.

Rees, W. E. (2002). An ecological economics perspective on sustainability and prospects for ending poverty. *Population and Environment, 24*(1), 15–46.

Reid, D. (1995). Sustainable development. In *An introductory guide* (First). Earthscan Publications Ltd.

Roodman, D. M. (2002). Trade slows. In Worldwatch Institute (Ed.), *Vital Signs 2002. The trends that are shaping our future: Vol. W.W. Norto*. Island Press.

Stavins, R. N., Wagner, A. F., & Wagner, G. (2003). Interpreting sustainability in economic terms: Dynamic efficiency plus intergenerational equity. *Economic Letters, 79*, 339–343.

UN. (1994). *International convention to combat desertification in countries experiences serious drought and/ or desertification particularly in Africa*. United Nations. http://www.unccd.int/convention/text/pdf/conv-eng.pdf

UN. (1997). *UN conference on environment and development (1992)*. United Nations. http://www.un.org/geninfo/bp/enviro.html

UN. (2002). *World summit on sustainable development plan of implementation*. United Nations. http://www.johannesburgsummit.org/html/documents/summit_docs/0409_plan_final.pdf

UN. (2005). *UN Millennium Development Goals*. United Nations. http://www.un.org/millenniumgoals/

UNCCD. (2021). The United Nations decade for deserts (2010–2020) and the fight against Desertification.

UNEP. (1972). *Declaration of the United Nations conference on the human Environment*. UNEP. http://www.unep.org/Documents/Default.asp?DocumentID=97&ArticleID=1503
United Nations. (2018). Sustainable development goals. http://www.un.org/sustainabledevelopment/sustainable-development-goals/
van Goethem, T., & van Landen, J. L. (2021). Biodiversity trends in a historical perspective. In *How Was Life? Volume II New Perspectives on Well-being and Global Inequality since 1820*. OECD. https://www.oecd-ilibrary.org/sites/2c94883d-en/index.html?itemId=/content/component/2c94883d-en#section-d1e66477
von Carlowitz, H. C. (2013). In J. Hamberger (Ed.), *Sylvicultura oeconomica*. Oekom.
Vos, R. O. (2007). Defining sustainability: A conceptual orientation. *Journal of Chemical Technology & Biotechnology, 82*(4), 334–339. https://doi.org/10.1002/jctb.1675
WCED. (1987). *Our common future* (First). Oxford University Press.
World Bank. (2021). *World development indicators*. World Bank.
Zadek, S. (1999). Stalking sustainability. *GMI, 26*, 1–11.

Chapter 3
Organisations and Their Contribution to Sustainability

3.1 Introduction

Organisations are an integral part of modern societies (Scott & Davis, 2015). They are also subsystems of a larger environmental system (Porter et al., 1975; Stacey, 1993) and have boundaries delineating the organisation's "inside" and "outside" (Scott & Davis, 2015). Organisations provide jobs in society (Pfeffer, 2010); they represent new types of social forms that have rights, capacities, and resources that are independent of those of its individual members (Scott & Davis, 2015) and have responsibilities to stakeholders (Freeman, 1984; Hill & Jones, 2001).

Organisations are complex social systems with sets of interrelated units engaged in joint problem-solving to accomplish a goal or objective (Jones, 2013; Porter et al., 1975; Rogers, 1962) and to create value and acceptable outcomes for stakeholders (Jones, 2013). Organisations have inextricably multiple, non-linear, connected processes, units, values, norms, behaviours, groups and individuals, affecting, and being affected by, each other, with myriad balancing feedback processes (Cyert & March, 1963; Kotter, 1996; Senge, 1999).

Organisations are, usually, divided into, as discussed by Holliday, Schmidheiny, and Watts (2002), civil society organisations (CSOs) (including society at large, education institutions, and non-governmental organisations), corporations (small, medium, large, national, international, and transnational), and public sector organisations (PSOs) (local, regional, and national). There are other types of organisations, e.g. hybrid organisations and public-private partnerships.

Organisations are affected by internal and external forces and conditions (Gebauer & Worch, 2015; Jones, 2013). Some authors (Kanter, 1999; Porter et al., 1975; Stacey, 1993) consider organisations as open systems, i.e. they exchange resources (e.g. energy, materials, labour, money, and information) with other organisations, while other authors (Drury & Farhoomand, 1999; Litvin, 2003) have considered them to be stand-alone units or islands or closed systems. However,

R. Lozano, *Toward Sustainable Organisations*, Strategies for Sustainability,
https://doi.org/10.1007/978-3-030-99676-5_3

organisations are semi-open (or semi-closed) systems, where some resources enter (e.g. employees when they arrive to work, raw materials, and energy), some resources exit (e.g. emissions and effluents, waste energy, products and by-products, employees when their work is finished), and some remain in the system (e.g. infrastructure, patents, organisation secrets, intellectual property, and organisational routines and behaviours) (Lozano, 2008a). The elements, structure, interdependences, interactions, and interconnectedness within an organisation and with other organisations, as well as the importance of boundaries between parts of an organisation and between organisations, and the roles of individuals within and across the boundaries can be understood with the help of systems theory (Stacey, 1993).

The study of systems, or systems thinking, can help to tackle complex organisational systems by (1) seeing wholes, their interrelations instead of linear cause-effect chains, and (2) understanding patterns of change, where there is "no right answer" for dealing with complexity (Senge, 1999). In particular, for organisations, systems theory helps to understand the interdependences, interactions and the interconnectedness of an organisation, and, among organisations, the importance of boundaries between parts of an organisation and between organisations and the roles of individuals within and across the boundaries (Stacey, 1993). It also helps to understand certain elements of the change process, such as leverage (Maurer, 1996; Senge, 1999), and stability or "steady state", referring to the capability of a system to return or remain in equilibrium after perturbations (Ludwig et al., 1997; McCann, 2000; Senge, 1999). It should be noted that systems do not necessarily behave as the sum of their individual parts, nor do the parts' behaviours necessarily follow the predicted behaviour of the whole (Hodge et al., 1999). However, in social systems, such as human organisations, the structures, norms, and interactions mould, through the sharing of mental models, the behavioural patterns of its members, tending to produce similar results, even from different individuals (Senge, 1999; Stacey, 1993).

3.2 Organisational System Elements

An organisation is composed of system elements (differentiated functions or groups) that are, in turn, composed of individuals (Luthans, 2002; Porter et al., 1975; Stacey, 1993); see Fig. 3.1. Such system elements are a key element of organisations (Luthans, 2002; Stacey, 1993). They have access to resources from which members of other groups are excluded (Dasgupta & Kanbur, 2003). Groups exist for a reason, and not just as arbitrary partitions of the system, including (Stacey, 1993) to accomplish complex interdependent tasks beyond the ability of individuals working alone, solve complex problems that require many inputs, provide a means of coordinating activities, facilitate implementation through generating participation and commitment, generate new ideas and creative solutions within the paradigm, and provide the opportunity for social interaction that improves morale and motivation. In large organisations, these are generally divided into top management, middle

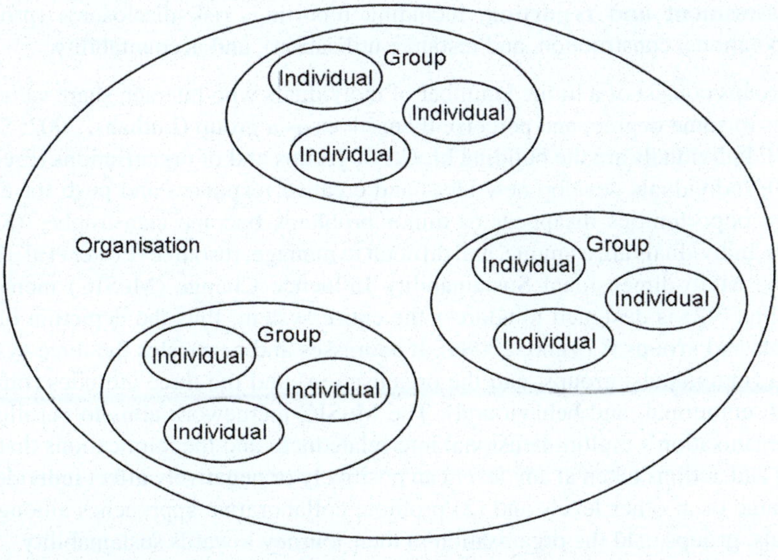

Fig. 3.1 The organisation as a social system

management, lower management, and rank-and-file positions. In small organisations, groups and their influence on the organisation tend to be diffuse.

Lozano and colleagues (see Lozano, 2012b; Lozano et al., 2017; Lozano, 2018) proposed the following parts for the system, in the context of sustainability (depending on the nature of the corporation, research and development may be part of operations and production, management and strategy, or organisational systems):

- **Operations and production**, including technologies, materials, energy sources, and product development, with closed-loop manufacturing combined with resource efficiency and effectiveness.
- **Management and strategy**, including productivity, investment and profit, business values and attitudes, objectives, vision and mission, strategies, products, and programmes, transparency and ethics in corporate decision-making, and stakeholder engagement, participation, and management.
- **Organisational systems**, such as people, culture, leadership styles, management skills and learning, problem-solving approaches, structures, systems, human resources and development, change management, and innovation.
- **Governance**, including ownership, the board of directors, institutional framework, and policies.
- **Supply chains**, considered as the outbound and inbound logistics and transportation, the customers, and consumers, including purchasing, advertising, promotional policies, procedures, and related activities.
- **Service provision**, including education and training, research and development, and customer service.

- **Assessment and reporting**, including reporting, risk disclosures (products, operations, construction, and resource utilisation), and accountability.

Groups consist of a limited number of individuals who interact, share values and norms to some degree, and perceive themselves as a group (Luthans, 2002; Stacey, 1993). Individuals are the building blocks of groups and of organisations (Freeman, 1984). Individuals develop new ideas and creative responses and push for change before opportunities disappear or minor problems become catastrophes (Kanter, 1999). Individuals are a unique, and difficult to manage, resource (Porter et al., 1975).

The Multi-dimensional Sustainability Influence Change (MuSIC) memework (see Fig. 3.2) is designed to address the entire system, thus the depiction of individuals and groups (Lozano, 2008a). It addresses and permeates the three system's levels (individuals, groups, and the organisation) and the three attitudes (informational, emotional, and behavioural). The MuSIC memework aims to (1) illustrate the organisation's multidimensional interrelatedness and the implications that decisions and actions taken at any level can positively or negatively affect individuals at the same or at other levels and (2) promote collaborative approaches among individuals, groups, and the organisation in their journey towards sustainability.

The MuSIC memework is built on two axes. The horizontal axis focuses on the sustainability internalisation and congruence between informational, emotional, and behavioural attitudes (i.e. knowledge, feelings and thoughts, and actions). An important point to note is that there has to be congruence among the three, without which there will be no transformation. The vertical axes focus upon the alignment among the different organisational levels (individuals, groups, and the organisations). The squares in the MuSIC memework are used to present a particular

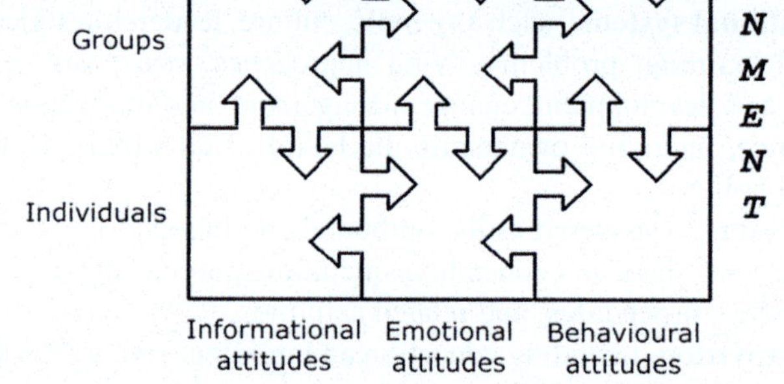

Fig. 3.2 Multi-dimensional Sustainability Influence Change (MuSIC) memework. Source: (Lozano, 2008a)

dimension (e.g. groups' emotional attitudes or individuals' informational attitudes). They do not depict equal size, influence, or power in the organisation. The arrows in the memework represent the different dimensions' relative influences and relations to the others. Congruence and alignment can help to better balance and calibrate the dimensions.

3.3 Organisation Stakeholders

No organisation is an island (Lozano & Huisingh, 2011); thus an organisation must balance the needs of multiple stakeholders (Freeman, 1984; Hill & Jones, 2001). Stakeholders can be internal (e.g. stockholders and employees, including management) and external (e.g. customers, suppliers, banks, environmentalists, and government, Argadoña, 1998; Freeman, 1984; Freeman et al., 2004). Stakeholders can also be divided into primary and secondary (Lindfelt, 2002; McIntosh et al., 1998). Table 3.1 shows examples of corporate stakeholders, where most of the stakeholder discourses have taken place. It should be noted that some of the stakeholder types might be different for the different organisation types.

The approaches and strategies on engaging with stakeholders can be grouped into six overarching typologies arising from differences in position and focus (Lozano, 2011) (which have been developed with a focus on corporations) (1) company responsibility choice; (2) benefits to stakeholders; (3) stakeholder emphasis; (4) stakeholder scope; (5) stakeholder influence strategies; and (6) corporate sustainability strategic influence typology.

The **company responsibility choice** typology is divided into three: (1) ethical, based on avoiding societal harms by fulfilling the corporation's ethical duties, i.e. the corporation is morally responsible to its stakeholders; (2) altruistic (or philanthropic), which demands that corporations help to alleviate societal problems, even if it reduces shareholder profits; and (3) strategic, based on benefiting society through activities that will bring benefits to the corporation in the long term. This categorisation approach is mainly an administrative exercise.

Table 3.1 Examples of primary, secondary, social, and non-social stakeholders of corporations. Source: (Lozano, 2011)

	Primary stakeholders	Secondary stakeholders
Social	• Shareholders (stockholders) and investors • Employees and managers • Customers • Unions • Suppliers and other business partners • Local communities	• Government and regulators • Civic institutions • Social pressure groups • The media and academia • Trade bodies • Competitors • General public
Non-social	• The natural environment • Future generations • Non-human species	• Environmental pressure groups • Animal welfare organisations

The **benefits to stakeholders** includes six types (Lozano, 2011): (1) harmonic, activities that clearly demonstrate benefit to shareholders in the long term and advance stakeholders', including societal, well-being. This is the most beneficial type; (2) mitigating, activities that mitigate stakeholder harm caused by the corporation, even without a legal responsibility, or clear benefits to shareholders; (3) philanthropic, activities which the corporation has no obligation to perform and which do not benefit shareholders, e.g. promoting AIDS prevention; (4) extrinsic, activities that reduce profits but improve stakeholder welfare; (5) pernicious, activities that increase profits but reduce stakeholder welfare; and (6) delusional, activities that reduce profits and stakeholder welfare. This categorisation provides a broader inclusion of different stakeholders than the corporate responsibility choice categorisation. However, it does not provide a detailed list, nor does it indicate what the motivations of the company might be to engage with sustainability, nor does it explicitly include the environment as a stakeholder.

The **stakeholder emphasis** typology is concerned with attention given to stakeholders (Lozano, 2011). It is subdivided into three categories: (1) inward-looking, dealing mainly with internal issues and stakeholders, such as human capital, health and safety, managing change, quality employment, lifelong learning, information, consultation and participation of workers, equal opportunities, integration of people with disabilities, anticipation of industrial change, and restructuring (Levitt, 1958); (2) outward-looking, dealing mainly with external issues and stakeholders, i.e. involvement with local communities, business partners, suppliers, customers, public authorities, and non-governmental organisations (NGOs) (Avi-Yonah, 2005; C.E.C., 2001; Waddock & Bodwell, 2007); such efforts are usually linked to reputation; and (3) broad perspective, dealing with both internal and external stakeholders (Frehs, 2003; Holme & Watts, 2000; Kaku, 2003). This typology does not include the motivation to engage with sustainability or explain who should be responsible and accountable within an organisation, and it omits the environment as an explicit stakeholder.

The **stakeholder scope** typology is divided into three categories: (1) philanthropy, which considers a corporation's responsibility to the community, where it operates or is located, to be fulfilled by charitable efforts (M. E. Porter & Kramer, 2003; Smith, 2003); (2) mainly dealing with social aspects, where the company aims to maximise its long-term profits (Fukukawa & Moon, 2004; Laffer et al., 2004; The Economist, 2005); and (3) addressing environmental and social aspects, where companies integrate those social and environmental concerns into their operations, with participative stakeholder engagement, deemed to be beneficial for the company's long-term interests (Holliday et al., 2002; Holme & Watts, 2000). Although this typology appears to be the most inclusive, it does not indicate why a corporation should engage with sustainability or who in the corporation should be responsible and accountable for such efforts.

The **corporate sustainability** strategic influence typology is aimed at providing a starting point to help managers identify how they are influencing, and where they could better influence, the different stakeholders, making their motivations explicit in doing so (Lozano, 2011), with seven options (the benefits to stakeholders from

Table 3.2 Corporate sustainability strategic influence typology. Source: (Lozano, 2011)

Strategy	Influences			
	Profit generation	Environment	Employees	External stakeholders
Holistic	Beneficial	Beneficial	Beneficial	Beneficial
Anthropocentric	Beneficial	Neutral	Beneficial	Beneficial
Insular	Beneficial	Neutral	Beneficial	Neutral
Green indulgences	Beneficial	Beneficial	Neutral	Neutral
Philanthropic	Beneficial	Neutral	Neutral	Beneficial
Fordian	Beneficial	Neutral	Negative	Neutral
Enronian	Negative	Negative	Negative	Negative

Cell coding
- Beneficial influences
- Neutral influences
- Negative influences

each option are presented in Table 3.2): (1) the holistic type generates profits and benefits the environment, the company's employees, and its external stakeholders; (2) the anthropocentric type generates profits and improves the welfare of all social stakeholders (internal and external), but does not consider influences on the environment; (3) the insular type refers to activities that focus only on improving profits and employees' welfare, where the effects on external stakeholders and the environment are neutral; (4) the green indulgences[1] type focuses on improving profits and the welfare of the environment, where the effects on employees and external stakeholders are neutral; (5) the philanthropic type includes those activities aimed at improving the welfare of external stakeholders, such as the communities where the company operates, where the effects on employees and the environment are neutral; (6) the Fordian type (following Henry Ford, who used brute force to break picket lines) refers to a focus solely on maximising shareholders' profits, but this ends up being detrimental to the welfare of employees; and (7) the Enronian type refers to those company activities that reduce shareholders profits, the welfare of employees, and external stakeholders and have negative impacts on the environment. These activities promote the profit of the few, in general the ones managing the corporation, by misleading/cheating the other stakeholders. From these seven strategies, the holistic type has the most benefits for stakeholders, whereas the Fordian and Enronian have the least benefits.

Some of the challenges in addressing stakeholders include the following: it is difficult to recognise and differentiate between stakeholders (Buchholz & Rosenthal, 2005; Langtry, 1994) and to meet the expectations of all stakeholder groups simultaneously (Argadoña, 1998; Dyllick & Hockerts, 2002); the inclusion of the natural

[1] "Indulgences" follows the medieval concept, which had the function of remitting time to be served by a sinner in purgatory by paying money to the representatives of the church (Goodin, 1994). Indulgences, in the sustainability context should be understood as providing money through charitable activities for the improvement of social welfare (Lozano, 2011).

environment, which lacks many of the characteristics of conventional stakeholders, e.g. specific identity and conscious decision-making capacities (Mitchel et al., 1997); and the consideration of a large number of stakeholders (Argadoña, 1998; Coelho et al., 2003).

3.4 Organisational System Interactions

In any organisation it is possible to find the following types of interactions (Lozano, 2008a):

- *Interpersonal interactions,* which take place between individuals, whether from the same group, different groups, the same organisation, or even from different organisations. This is the simplest one.
- *Intergroup interactions,* which are found among groups within a particular organisation or from different ones. It should be noted that even if the agent of these interactions is an individual, there is a difference between interpersonal and intergroup interactions. In the former the relation is on a personal level, while in the latter the relation deals with topics of concern to both groups, with agents acting on behalf of their respective groups. It is also possible to find interactions between a particular individual and a group.
- *Inter-organisational interactions,* which refers to interactions among different organisations.

An example of interpersonal relations with an influential individual is presented in Fig. 3.3, where it can be observed that individual 2 (I2) has strong links with the others, making him/her an influential person in the group. It can also be observed that individual 5 (I5) has weak links to the other individuals; thus his/her influence in the system is low.

An example of interactions among groups in an organisation is presented in Fig. 3.4. This figure presents an ideal case where the relations, influence, and

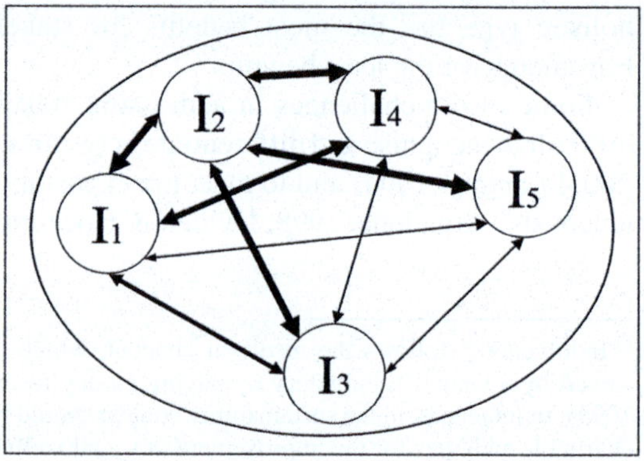

Fig. 3.3 Individual interactions within a group. Source: (Lozano, 2008a)

Fig. 3.4 Group interactions within an organisation. Source: (Lozano, 2008a)

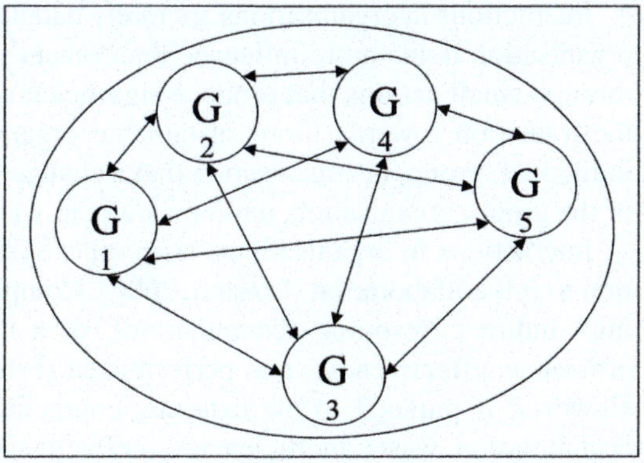

Table 3.3 Interactions in the organisational system, among individuals, groups, and organisations

	Individuals	Groups	Organisation
Individuals	Inter-personal: among individuals within the same group or different groups in the organisation (Luthans, 2002)	Intra-group (I-G): the individual serving as an agent to interact with another group (Stacey, 1993)	Intra-organisational (I-O): from the individuals to the organisation (Andersson et al., 2005; Carr, 2001; Porter et al., 1975)
Groups	Intra-group (G-I): the group serving as an agent to interact with a particular individual (Stacey, 1993)	Inter-group: from groups within the same organisation (Freeman, 1984; Porter et al., 1975)	Intra-organisational (G-O): from the groups to the organisation (Lozano, 2008a)
Organisation	Intra-organisational (O-I): from the organisations to individuals ((Andersson et al., 2005; Carr, 2001; Porter et al., 1975)	Intra-organisational (O-G): from the organisation to groups (Lozano, 2008a)	Inter-organisational: interaction with the external environment (Lozano, 2008a)

communication within the groups are balanced, and all can work collaboratively to achieve the optimum outcome. Group interactions are similar to individual interactions. It should be noted that in many cases, rival groups vie to keep the majority, or all, of the available resources or demand a dominant position in the organisation, which leads to internal conflicts in the organisation.

Interactions can be of the following forms (see Table 3.3): (1) reciprocal interactions between the individual and the organisation, exchanging values and cultural substance (Andersson et al., 2005; Carr, 2001; Porter et al., 1975); (2) between individuals and groups, which is characteristic of leaders and groups, but it is also true of relations among members of the group (Stacey, 1993); (3) between groups (Freeman, 1984; Porter et al., 1975); and (4) between individuals from the same or different groups (Luthans, 2002). Additionally, there can also be interactions between groups and the organisation and between organisations (Lozano, 2008a). These interactions are presented in Table 3.3.

Interactions in organisations are rarely balanced. Usually an individual, group, or organisation holds more influence than others. Influence can be used for leverage, through small actions that produce significant and enduring improvements, to help the transition towards more sustainable organisations. Convincing an influential individual, group, or organisation that a change is required eases the transformation of the group, organisation, and/or society to which they belong (Maurer, 1996).

Interactions in organisations, as in other systems, range from extreme competition to full collaboration (Lozano, 2006). Competition, whether direct (power seeking), indirect (resource accumulation), or a mixture of both, can help improve efficiency, effectiveness, and performance (Friedman, 1970; Hill & Jones, 2001). However, if pursued to the extreme, competition can lead to excessive resource accumulation, waste, inefficiencies, ineffectiveness, poor performance, and creation of monopolies and oligopolies (Arora et al., 1998; Parente & Prescott, 1999). Such extremes, or pure forms, affect all types of interactions. Extreme competition often works against the best interests of the organisation in the long term (Porter et al., 1975), since it can lead to conflicts (Carley & Christie, 2000; Lozano, 2007).

Collaboration is used interchangeably with communication, coordination, and cooperation (Dawson, 1994; Denise, 1999; Lozano, 2007, 2008a). However, these terms are fundamentally different. Coordination refers to activities carried out by different individuals in order to make them compatible for a common purpose or result (Chilosi, 2003; Denise, 1999; Lozano, 2008a). Cooperation refers to engaging in work on monitoring and evaluation, learning from each other, and sharing experiences (Lozano, 2008a; Tunstall, 2005).

Collaboration refers to using information to create something new, seeking divergent insights and spontaneity, jointly developing proposals, sharing information, planning joint workshops, and raising funds together, among other activities (Lozano, 2008a). It thrives on differences and dissent (Chilosi, 2003; Denise, 1999; Lozano, 2008a). Collaboration results in benefits from differences in knowledge and perspectives, natural learning, working in cross-functional or inter-organisational teams, developing new understandings, designing new products, and solving problems while, at the same time, offering benefits to all those involved in the process (Chilosi, 2003; Denise, 1999; Fadeeva, 2004; Kaku, 2003; Lozano, 2008a). It relies on sharing and openness (Bevan et al., 2012) by using information, divergent insights, and spontaneity, (Denise, 1999).

Two key concepts in collaboration are alignment and congruence. Alignment is when key organisational performance influence factors (e.g. understanding, common orientation, values and shared priorities, goals strategies, and employees) are shared and send the same message (Bovey & Hede, 2001; Collins & Porras, 2002; Gill, 2003; Lozano, 2008a). Congruence refers to consistency between informational (what is learnt), emotional (what is thought), and behavioural attitudes (what is done) (Lozano, 2008a).

Collaboration can optimise financial and human capital, ease access to markets and knowledge, enrich creativity, and avoid confrontation (Fadeeva, 2004). It can help to catalyse innovation, research and development, policy development, and implementation, reduce conflicts (Langer & Schön, 2003), help organisations to survive, and go beyond mere survival, in the long term (Lozano, 2007, 2008a). Collaboration can also help organisations develop products and services (Camarinha-Matos et al., 2009); however, it may lead to less efficient decision-making and conflicts over resources and technical issues (Troy et al., 2008). It, usually, provides more benefits to organisations than when they work on their own (Allred et al., 2011). Collaboration is the sine qua non of effective, planned change (Bennis et al., 1969; Lozano, 2008a).

Collaboration plays a key role in helping organisations become more sustainable (Govindan et al., 2016; Lozano, 2007, 2008a; Seuring & Gold, 2013), through including developing new technologies (Kishna et al., 2017) and business models (Witjes & Lozano, 2016), improving recovery (reducing, reusing, and recycling resources) (Lozano et al., 2013), fostering innovation (McLachlan et al., 2015), increasing research and education (Caniglia et al., 2017), and linking to the community (Murray & Zautra, 2012).

The challenges that organisations face from collaboration in becoming more sustainable include information (referring to who benefits, and the real, or hidden, agenda) (Graci, 2013), bargaining (how to split the gains) (Lozano, 2007), freeriding (where those who choose not to participate still get the benefits) (Chilosi, 2003; Vaaland, 2004), coordination problems (Steensma, 1996), risk of jeopardising important and unique resources (Genefke, 2000; Grimpe & Kaiser, 2010), conflicting needs (Lozano, 2007; Walsh & Maloney, 2007), relationship conflicts (Kumar & Van Dissel, 1996), data conflicts (Hertzfeld et al., 2006; Zafar & Kantola, 2019), value conflicts (Nudurupati et al., 2011), structural conflicts (Vaaland, 2004), and operational dependencies between the activities of the different actors (Noll et al., 2010).

The "organisational sustainability collaboration" framework, as illustrated in Fig. 3.5, shows the five stages of collaboration dependent on the factors

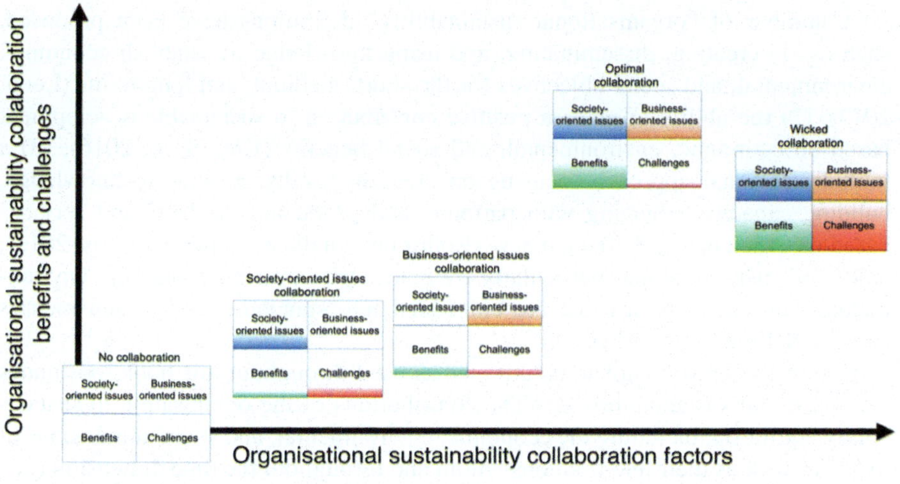

Fig. 3.5 Organisational sustainability collaboration framework, where blue indicates an increase in society-oriented issues, orange for business-oriented issues, green the benefits, and red the challenges. Source: (Lozano et al., 2021)

(society-oriented issues and business-oriented issues) and the benefits and challenges obtained from collaborating. The figure shows that the best alternative is "optimal collaboration", where the factors are almost equal for benefits, with few challenges. The figure also shows that when there are no limits to collaboration, then the challenges increase considerably (Lozano et al., 2021).

3.5 Organisations' Sustainability Efforts

Organisations have been instrumental in driving sustainability (Danter et al., 2000; Holliday et al., 2002; Jennings & Zandbergen, 1995), where the importance of sustainability's dimensions depends on an organisation's nature and purpose (Soyka, 2012). The impacts of organisations have been highest on the social dimension (67%), followed by the economic (61%), and then the environmental (55%) (Lozano, 2018).

The last decade has seen an increasing interest in organisational sustainability (Corsi & Arru, 2020; Hörisch et al., 2015; Lozano, 2008a, 2018; Pfeffer, 2010). Some examples of organisational sustainability include individuals-organisation-global environment relations (Edwards, 2009), organisational culture within the context of corporate sustainability (Linnenluecke et al., 2009; Linnenluecke & Griffiths, 2010), initiatives to embed sustainability into companies (Lozano, 2012b; Robèrt, 2000; Robèrt et al., 2002), organisational change management for sustainability (Lozano, 2013, 2015), recovery efforts through circular economy (Barreiro-Gen & Lozano, 2020), governance (Temel et al., 2021), and links between assessment and reporting and organisational change management (Ceulemans et al., 2015; Domingues et al., 2017; Lozano et al., 2016).

A number of "organisational sustainability" definitions have been proposed, such as (1) creating, disseminating, and using knowledge to establish economic, environmental, and social objectives for the short, medium, and long terms (Leon, 2013); (2) the ability to make a positive contribution to sustainable development based on economic, environmental, and social benefits (Kim et al., 2016); (3) a multidimensional approach focusing on creating results, generating knowledge, building capacity, engaging with partners, and producing products and services based on efficiency and effectiveness (Rodríguez-Olalla & Avilés-Palacios, 2017); and (4) decision-making, particularly by managers, covering economic, environmental, and social issues that are embedded into standards, codes, and metrics (Batista & Francisco, 2018).

One of the most complete definitions, as used throughout this book, postulates organisational sustainability as: "The contributions of the organisation to sustainability equilibria, including the economic, environmental, and social dimensions of today, as well as their interrelations within and throughout the time dimension (i.e., the short-, long-, and longer-term)". This entails the continuous incorporation and implementation of sustainability issues in the organisation's system elements (operations and production, strategy and management, governance, organisational systems, service provision, and assessment and reporting), as well as change processes and their rate of change. The system elements and change processes transform the inputs (regarding materials and resources that have economic, environmental, and social value) into outputs (products, services, and waste, with their economic, environmental, and social value). These fulfil the organisation's goal or objective, based on resource efficiency and effectiveness. The organisation is affected by the organisation's non-human and human resources (i.e., individuals, groups, culture, values, attitudes, and norms), its infrastructure, its supply chain (upstream and downstream), and the interactions with its stakeholders (internal, interconnecting, and external)" (Lozano, 2018).

Figure 3.6 depicts the above definition with the following elements: (I) inputs, including material resources and energy, economic value, (intrinsic) environmental value, and human resources (that enter the company); (II) system elements (operations and production, management and strategy, organisational systems, governance, and assessment and reporting) and change processes, which link the inputs and outputs and generate waste; (III) stakeholders (internal, interconnecting, and external); (IV) outputs, including products and services, environmental value, and human resources (that exit the company); and (V) purpose fulfilment, based on resource efficiency. It should be noted that the organisation has tangible and intangible resources and part of the products and services may come back through recovery as material resources or energy. The colours in Fig. 3.6 indicate the different flow paths: white refers to material resources and energy transformed through the processes to products and services; red refers to the economic value of the input materials and their increase to output economic valued, through yellow, purpose fulfilment (based on resource efficiency and effectiveness); lime indicates the

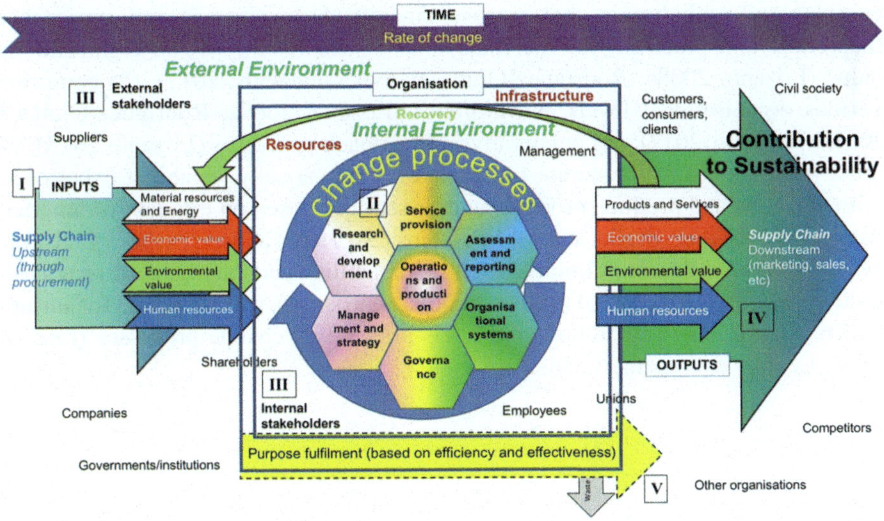

Fig. 3.6 Organisational sustainability framework. Source: (Lozano, 2018)

environmental value of the inputs and outputs, as well as recovery; blue refers to the human resources that enter and exit the company, as modified by the system elements and change; purple highlights the time dimension; green font indicates the environment (internal and external); blue font highlights the supply chain; red font indicates the resources and infrastructure; and grey shows the waste generated by the processes.

It should be noted that some authors use the term "sustainable organisation" (Leon, 2013; Rodríguez-Olalla & Avilés-Palacios, 2017), but this implies that the organisation is sustainable, which is not possible since sustainability is a dynamic ideal that can always be improved (Lozano, 2008b). The term "organisational sustainability", defined as the contributions of an organisation to sustainability, is much more encompassing since it includes how sustainability is integrated into its system elements and processes, and how the organisation contributes to sustainable development.

Most of the literature on organisational sustainability has focused on corporations (Dyllick & Hockerts, 2002; Pfeffer, 2010; Schaltegger & Burrit, 2005; Tschopp, 2005), followed by education institutions, in particular higher education (Dlouhá et al., 2013; Lozano et al., 2014). Limited, yet increasing, attention has been directed to PSOs and other CSOs (Domingues et al., 2017; Dumay et al., 2010; Guthrie & Farneti, 2008; Lodhia et al., 2012). There have been only a few papers comparing the types of organisations, e.g. Blanco-Portela et al. (2017), comparison of drivers for and barriers to change between higher education institutions (HEIs) and corporations. Most organisational sustainability efforts have focused on the "technocentric" (Doppelt, 2003; Linnenluecke et al., 2009) or managerial ploys (Henriques & Richardson, 2005). There has been increasing interest on organisational systems and change processes (Lozano, 2012a).

Incorporating, integrating, and implementing sustainability in organisations is a complex and multidimensional process, since it (1) involves several approaches, theoretical and practical, for the inclusion of sustainability as an organisational value (Lozano, 2008a; Rodríguez-Olalla & Avilés-Palacios, 2017); (2) comprises various organisational levels and attitudes (Lozano, 2008a; Rodríguez-Olalla & Avilés-Palacios, 2017); (3) entails myriad feedback processes (Lozano, 2015); (4) must maintain results, generate knowledge, build capacity, establish experiences with partners, and produce products and/or services based on efficiency and effectiveness (Rodríguez-Olalla & Avilés-Palacios, 2017); (5) addresses the organisation system elements, the four dimensions of sustainability, and the organisation's stakeholders (Lozano, 2012b); (6) the use of tools, approaches, and initiatives for sustainability is still under-researched; and (7) is affected by change processes (Lozano, 2013; Lozano & von Haartman, 2018).

References

Allred, C. R., Fawcett, S. E., Wallin, C., & Magnan, G. M. (2011). A dynamic collaboration capability as a source of competitive advantage. *Decision Sciences, 42*(1), 129–161. https://doi.org/10.1111/j.1540-5915.2010.00304.x

Andersson, L., Shivarajan, S., & Blau, G. (2005). Enacting ecological sustainability in the MNC: A test of an adapted value-belief-norm framework. *Journal of Business Ethics, 59*, 259–305.

Argadoña, A. (1998). The stakeholder theory and the common good. *Journal of Business Ethics, 17*, 1093–1102.

Arora, A., Landau, R., & Rosenberg, N. (1998). *Chemicals and long-term economic growth. Insights from the chemical indsutry*. John Wiley & Sons, Inc.

Avi-Yonah, R. (2005). The cyclical transformations of the corporate form: A historical perspective on corporate social responsibility. *Delaware Journal of Corporate Law, 30*, 767–818. http://law.bepress.com/umichlwps-olin/art38/

Barreiro-Gen, M., & Lozano, R. (2020). How circular is the circular economy? Analysing the implementation of circular economy in organisations. *Business Strategy and the Environment, 29*(8), 3484–3494. https://doi.org/10.1002/bse.2590

Batista, A. A. D. S., & Francisco, A. C. D. (2018). Organizational sustainability practices: A study of the firms listed by the corporate sustainability index. *Sustainability (Switzerland), 10*(1). https://doi.org/10.3390/su10010226

Bennis, W. G., Benne, K. D., & Chin, R. (1969). *The planning of change* (2nd ed.). Holt, Rinehart and Winston, Inc.

Blanco-Portela, N., Benayas, J., Pertierra, L. R. L. R., & Lozano, R. (2017). Towards the integration of sustainability in higher education institutions: A review of drivers of and barriers to organisational change and comparison against those found of companies. *Journal of Cleaner Production, 166*, 563–578. https://doi.org/10.1016/j.jclepro.2017.07.252

Bovey, W. H., & Hede, A. (2001). Resistance to organisational change: The role of defence mechanisms. *Journal of Managerial Psychology, 16*(7), 534–548.

Buchholz, R. A., & Rosenthal, S. B. (2005). Toward a contemporary conceptual framework for stakeholder theory. *Journal of Business Ethics, 58*, 137–148.

C.E.C. (2001). *Promoting a European framework for corporate social responsibility*. Commission of the European Communities.

Camarinha-Matos, L. M., Afsarmanesh, H., Galeano, N., & Molina, A. (2009). Collaborative networked organizations - concepts and practice in manufacturing enterprises. *Computers and Industrial Engineering, 57*(1), 46–60. https://doi.org/10.1016/j.cie.2008.11.024

Caniglia, G., Luederitz, C., Groß, M., Muhr, M., John, B., Withycombe Keeler, L., von Wehrden, H., Laubichler, M., Wiek, A., & Lang, D. (2017). Transnational collaboration for sustainability in higher education: Lessons from a systematic review. In *Journal of cleaner production* (Vol. 168, pp. 764–779). Elsevier Ltd.. https://doi.org/10.1016/j.jclepro.2017.07.256

Carley, M., & Christie, I. (2000). *Managing sustainable development* (2nd ed.). Earthscan Publications Ltd. www.earthscan.co.uk

Carr, A. (2001). Understanding emotion and emotionality in a process of change. *Journal of Organizational Change, 14*(5), 421–434.

Ceulemans, K., Lozano, R., & Alonso-Almeida, M. (2015). Sustainability reporting in higher education: Interconnecting the reporting process and Organisational change Management for Sustainability. *Sustainability, 7*(7), 8881–8903. https://doi.org/10.3390/su7078881

Chilosi, A. (2003). *Coordination, cooperation, and the extended Coasean approach to economic policy*. Facolta di Scienze Politiche.

Coelho, P. R. P., McClure, J. E., & Spry, J. A. (2003). The social responsibility of corporate management: A classical critique. *Mid - American Journal of Business, 18*(1), 15–24.

Collins, J., & Porras, J. I. (2002). Built to last. In *Successful habits of visionary companies*. Harper Business Essentials.

Corsi, K., & Arru, B. (2020). Role and implementation of sustainability management control tools: Critical aspects in the Italian context. *Accounting, Auditing and Accountability Journal, 34*(9), 29–56. https://doi.org/10.1108/AAAJ-02-2019-3887

Cyert, R. M., & March, J. G. (1963). *A behavioural theory of the firm*. Prentice-Hall.

Danter, K. J., Griest, D. L., Mullins, G. W., & Norland, E. (2000). Organizational change as a component of ecosystem management. *Society and Natural Resources, 13*(6), 537–547. https://doi.org/10.1080/08941920050114592

Dasgupta, I., & Kanbur, R. (2003). Community and anti-poverty targeting. *The Journal of Economic Inequality, 3*(3), 281–302.

Dawson, P. (1994). *Organizational change. A processual approach*. Paul Chapman Publishing Ltd.

Denise, L. (1999). Collaboration vs. C-three (cooperation, coordination, and communication). *Innovations, 7*(3).

Dlouhá, J., Huisingh, D., & Barton, A. (2013). Learning networks in higher education: Universities in search of making effective regional impacts. *Journal of Cleaner Production, 49*, 5–10. https://doi.org/10.1016/j.jclepro.2013.01.034

Domingues, A. R., Lozano, R., Ceulemans, K., & Ramos, T. B. (2017). Sustainability reporting in public sector organisations: Exploring the relation between the reporting process and organisational change management for sustainability. *Journal of Environmental Management, 192*, 292–301. https://doi.org/10.1016/j.jenvman.2017.01.074

Doppelt, B. (2003). Leading change toward sustainability. In *A change-management guide for business, government and civil society*. Greenleaf Publishing.

Drury, D. H., & Farhoomand, A. (1999). Innovation diffusion and implementation. *International Journal of Innovation Management, 3*(2), 133–157.

Dumay, J., Guthrie, J., & Farneti, F. (2010). Gri sustainability reporting guidelines for public and third sector organizations. In *Public Management Review, November 2012* (pp. 37–41).

Dyllick, T., & Hockerts, K. (2002). Beyond the business case for corporate sustainability. *Business Strategy and the Environment, 11*, 130–141.

Edwards, M. (2009). *Organisational transformation for sustainability: An integral metatheory*. Routledge.

Fadeeva, Z. (2004). Promise of sustainability collaboration - potential fulfilled? *Journal of Cleaner Production, 13*, 165–174.

Freeman, R. E. (1984). Strategic management. A stakeholder approach. Pitman Publishing Inc.

Freeman, R. E., Wicks, A. C., & Parmar, B. (2004). Stakeholder theory and "The Corporate Objective Revisited". *Organization Science, 15*(3), 364–369.

Frehs, J. (2003). *Corporate social responsibility: Lessons learned*. Natural Resources Canada.

Friedman, M. (1970). *The social responsibility of business is to increase its profits*. The New York Times Magazine. September 13, 1970.

Fukukawa, K., & Moon, J. (2004). A Japanese model of corporate social responsibility? *Journal of Corporate Citizenship, 16*, 45–59.

Gebauer, H., & Worch, H. (2015). Absorptive capacity (of Organizations). In *International Encyclopedia of the Social & Behavioral Sciences* (Vol. 1, 2nd ed.). Elsevier. https://doi.org/10.1016/B978-0-08-097086-8.73106-0

Genefke, J. (2000). *Collaboration costs!* University of Aarhus. Department of Management.

Gill, R. (2003). Change management - or change leadership? *Journal of Change Management, 3*(4), 307–318.

Govindan, K., Seuring, S., Zhu, Q., & Azevedo, S. G. (2016). Accelerating the transition towards sustainability dynamics into supply chain relationship management and governance structures. *Journal of Cleaner Production*. https://doi.org/10.1016/j.jclepro.2015.11.084

Graci, S. (2013). Collaboration and partnership development for sustainable tourism. *Tourism Geographies, 15*(1), 25–42. https://doi.org/10.1080/14616688.2012.675513

Grimpe, C., & Kaiser, U. (2010). Balancing internal and external knowledge acquisition: The Gains and pains from R&D outsourcing. *Journal of Management Studies, 47*(8), 1483–1509. https://doi.org/10.1111/j.1467-6486.2010.00946.x

Guthrie, J., & Farneti, F. (2008). GRI sustainability reporting by Australian public sector organizations. *Public Money & Management, 28*(6), 361–366.

Henriques, A., & Richardson, J. (2005). *The triple bottom line. Does it all add up?* Earthscan.

Hertzfeld, H. R., Link, A. N., & Vonortas, N. S. (2006). Intellectual property protection mechanisms in research partnerships. *Research Policy, 35*(6), 825–838. https://doi.org/10.1016/j.respol.2006.04.006

Hill, C. W. L., & Jones, G. R. (2001). Strategic management: An integrated approach. In *Strategic Management An Integrated Approach* (Fifth). Houghton Mifflin Company.

Hodge, R. A., Hardi, P., & Bell, D. V. J. (1999). *Seeing change through the lens of sustainability.* The International Institute for Sustainable Development.

Holliday, C. O. J., Schmidheiny, S., & Watts, P. (2002). Walking the talk. In *The Business Case for Sustainable Development.* Greenleaf Publishing.

Holme, R., & Watts, P. (2000). *Corporate social responsibility: Making good business sense.* WBCSD.

Hörisch, J., Ortas, E., Schaltegger, S., & Álvarez, I. (2015). Environmental effects of sustainability management tools: An empirical analysis of large companies. *Ecological Economics, 120,* 241–249. https://doi.org/10.1016/j.ecolecon.2015.11.002

Jennings, P. D., & Zandbergen, P. A. (1995). Ecologically sustainable organizations: An institutional approach. *Academy of Management Review, 20*(4), 1015–1052. https://doi.org/10.5465/AMR.1995.9512280034

Jones, G. R. (2013). *Organizational theory, design, and change* (Seventh). Pearson Education Limited.

Kaku, R. (2003). The path of Kyosei. In *Harvard Business Review on Corporate Responsibility* (Issue July–August). Harvard Business School Press.

Kanter, R. M. (1999). *The change masters* (first). International Thomson Business Press.

Kim, W., Khan, G. F., Wood, J., & Mahmood, M. T. (2016). Employee engagement for sustainable organizations: Keyword analysis using social network analysis and burst detection approach. *Sustainability (Switzerland), 8*(7), 1–11. https://doi.org/10.3390/su8070631

Kishna, M., Niesten, E., Negro, S., & Hekkert, M. P. (2017). The role of alliances in creating legitimacy of sustainable technologies: A study on the field of bio-plastics. *Journal of Cleaner Production.* https://doi.org/10.1016/j.jclepro.2016.06.089

Kotter, J. P. (1996). *Leading change.* Harvard Business School Press.

Kumar, K., & Van Dissel, H. G. (1996). Sustainable collaboration: Managing conflict and cooperation in interorganizational systems. *MIS Quarterly: Management Information Systems, 20*(3), 279–299. https://doi.org/10.2307/249657

Laffer, A. B., Coors, A., & Winegarden, W. (2004). *Does corporate social responsibility enhance business profitability?* Laffer Associates.

Langer, M. E., & Schön, A. (2003). Enhancing corporate sustainability. A framework based evaluation tools for sustainable development. Forschungsschwerpunkt Nachhaltigkeit und Umweltmanagement, Wirtschaftsuniversität Wien.

Langtry, B. (1994). Stakeholders and the moral responsibilities of business. *Business Ethics Quarterly, 4*(4).

Leon, R. D. (2013). The sustainable knowledge-based organisation from a dynamic capabilities approach. *International Journal of Learning and Intellectual Capital, 10*(3/4), 243. https://doi.org/10.1504/IJLIC.2013.057433

Levitt, T. (1958). *The dangers of social responsibility* (pp. 41–50). Harvard Business Review.

Lindfelt, L.-L. (2002). *Corporate social responsibility in the new global economy.* Företagsekonomiska Institutionen Uppsala Universitet.

Linnenluecke, M. K., & Griffiths, A. (2010). Corporate sustainability and organizational culture. *Journal of World Business, 45*(4), 357–366. https://doi.org/10.1016/j.jwb.2009.08.006

Linnenluecke, M. K., Russell, S. V., Griffi, A., Linnenluecke, M. K., & Griffiths, A. (2009). Subcultures and sustainability practices : The impact on understanding corporate sustainability. *Business Strategy and the Environment, 18*(December 2007), 432–452. https://doi.org/10.1002/bse

Litvin, D. (2003). *Empires of profit. Commerce, conquest and corporate responsibility*. TEXERE Publishing Limited.

Lodhia, S., Jacobs, K., & Park, Y. J. (2012). Driving public sector environmental reporting. *Public Management Review, 14*(5), 631–647.

Lozano, R. (2006). Collaboration as pathway for sustainability. In *Environmental management for sustainable universities*.

Lozano, R. (2007). Collaboration as a pathway for sustainability. *Sustainable Development, 381*(March), 370–381. https://doi.org/10.1002/sd.322

Lozano, R. (2008a). Developing collaborative and sustainable organisations. *Journal of Cleaner Production, 16*(4), 499–509. https://doi.org/10.1016/j.jclepro.2007.01.002

Lozano, R. (2008b). Envisioning sustainability three-dimensionally. *Journal of Cleaner Production, 16*(17), 1838–1846. https://doi.org/10.1016/j.jclepro.2008.02.008

Lozano, R. (2011). Addressing stakeholders and better contributing to sustainability through game theory. *Journal of Corporate Citizenship, 43*(0), 44–62.

Lozano, R. (2012a). Orchestrating Organisational changes for corporate sustainability. *Greener Management International, 57*, 43–64. https://doi.org/10.9774/GLEAF.3062.2007.sp.00005

Lozano, R. (2012b). Towards better embedding sustainability into companies' systems: An analysis of voluntary corporate initiatives. *Journal of Cleaner Production, 25*(0), 14–26. https://doi.org/10.1016/j.jclepro.2011.11.060

Lozano, R. (2013). Are companies planning their Organisational changes for corporate sustainability? An analysis of three case studies on resistance to change and their strategies to overcome it. *Corporate Social Responsibility and Environmental Management, 20*(5), 275–295. https://doi.org/10.1002/csr.1290

Lozano, R. (2015). A holistic perspective on corporate sustainability drivers. *Corporate Social Responsibility and Environmental Management, 22*(1), 32–44. https://doi.org/10.1002/csr.1325

Lozano, R. (2018). Proposing a definition and a framework of Organisational sustainability: A review of efforts and a survey of approaches to change. *Sustainability, 10*(4), 1157. https://doi.org/10.3390/su10041157

Lozano, R., Barreiro-Gen, M., & Zafar, A. (2021). Collaboration for organizational sustainability limits to growth: Developing a factors, benefits, and challenges framework. *Sustainable development, July 2020*, sd.2170. https://doi.org/10.1002/sd.2170.

Lozano, R., Carpenter, A., & Satric, V. (2013). Fostering green chemistry through a collaborative business model: A chemical leasing case study from Serbia. *Resources, Conservation and Recycling, 78*(November), 136–144. https://doi.org/10.1016/j.resconrec.2013.07.007

Lozano, R., Ceulemans, K., Alonso-Almeida, M., Huisingh, D., Lozano, F. J. F. J., Waas, T., Lambrechts, W., Lukman, R., & Hugé, J. (2014). A review of commitment and implementation of sustainable development in higher education: Results from a worldwide survey. *Journal of Cleaner Production, 108*. https://doi.org/10.1016/j.jclepro.2014.09.048

Lozano, R., & Huisingh, D. (2011). Inter-linking issues and aspects in sustainability reporting. *Journal of Cleaner Production, 19*(2–3), 99–107. https://doi.org/10.1016/j.jclepro.2010.01.004

Lozano, R., Nummert, B., & Ceulemans, K. (2016). Elucidating the relationship between sustainability reporting and organisational change management for sustainability. *Journal of Cleaner Production, 125*, 168–188. https://doi.org/10.1016/j.jclepro.2016.03.021

Lozano, R., Suzuki, M., Carpenter, A., & Tyunina, O. (2017). An analysis of the contribution of Japanese business terms to corporate sustainability: Learnings from the "looking-glass" of the east. *Sustainability (Switzerland), 9*(2), 1–17. https://doi.org/10.3390/su9020188

Lozano, R., & von Haartman, R. (2018). Reinforcing the holistic perspective of sustainability: Analysis of the importance of sustainability drivers in organizations. *Corporate Social Responsibility and Environmental Management, 25*(4), 508–522. https://doi.org/10.1002/csr.1475

Ludwig, D., Walker, B., & Holling, C. S. (1997). Sustainability, stability, and resilience. In. *Conservation Ecology, 1*(1) http://www.ecologyandsociety.org/vol1/iss1/art7/

Luthans, F. (2002). *Organizational behavior*. McGraw-Hill.

Maurer, R. (1996). Beyond the wall of resistance. In *Unconventional strategies that build support for change*. Bard Book, Inc..

McCann, K. S. (2000). The diversity - stability debate. *Nature, 405*(May), 228–233.

McIntosh, M., Leipziger, D., Jones, K. K. K., & Coleman, G. (1998). Corporate citizenship. In *Successful strategies for responsible companies*. Financial Times. Pitman Publishing.

McLachlan, M., Hamann, R., Sayers, V., Kelly, C., & Drimie, S. (2015). Fostering innovation for sustainable food security: The southern Africa food lab. In *The Business of Social and Environmental Innovation* (pp. 163–181). Springer International Publishing. https://doi.org/10.1007/978-3-319-04051-6_9

Mitchel, R. K., Agle, B. R., & Wood, D. J. (1997). Toward a theory of stakeholder identification and salience: Defining the principle of who and what really counts. *Academy of Management Review, 22*(4), 853–886.

Murray, K., & Zautra, A. (2012). Community resilience: Fostering recovery, sustainability, and growth. In *The social ecology of resilience: A handbook of theory and practice* (pp. 337–345). Springer. https://doi.org/10.1007/978-1-4614-0586-3_26

Noll, J., Beecham, S., & Richardson, I. (2010). Global software development and collaboration: Barriers and solutions. *ACM Inroads, 1*(3), 66–78. https://doi.org/10.1145/1835428.1835445

Nudurupati, S. S., Bititci, U. S., Kumar, V., & Chan, F. T. S. (2011). State of the art literature review on performance measurement. *Computers and Industrial Engineering, 60*(2), 279–290. https://doi.org/10.1016/j.cie.2010.11.010

Parente, S. L., & Prescott, E. C. (1999). Monopoly rights: A barrier to riches. *The American Economic Review, 89*(5), 1216–1233.

Pfeffer, J. (2010). Building sustainable organizations: The human factor. *Academy of Management Perspectives, 24*, 34–45. https://doi.org/10.5465/AMP.2010.50304415

Porter, L. W., Lawler, E. E., & Hackman, J. R. (1975). *Behavior in organizations*. McGraw-Hill.

Porter, M. E., & Kramer, M. R. (2003). The competitive advantage of corporate philanthropy. In *Harvard business review on corporate responsibility*. Harvard Business School Press.

Robèrt, K.-H. (2000). Tools and concepts for sustainable development, how do they relate to a general framework for sustainable development, and to each other? *Journal of Cleaner Production, 8*(3), 243–254. https://doi.org/10.1016/S0959-6526(00)00011-1

Robèrt, K.-H., Schmidt-Bleek, B., Aloisi de Larderel, J., Basile, G., Jansen, J. L., Kuehr, R., Price Thomas, P., Suzuki, M., Hawken, P., & Wackernagel, M. (2002). Strategic sustainable development - selection, design and synergies of applied tools. *Journal of Cleaner Production, 10*, 197–214.

Rodríguez-Olalla, A., & Avilés-Palacios, C. (2017). Integrating sustainability in Organisations: An activity-based sustainability model. *Sustainability, 9*(12), 1072. https://doi.org/10.3390/su9061072

Rogers, E. M. (1962). *Diffusion of innovations* (4th ed.). The Free Press of Glencoe.

Schaltegger, S., & Burrit, R. L. (2005). Corporate sustainability. In H. Folmer & T. Tietenberg (Eds.), *The international yearbook of environmental and resource economics 2005/2006* (pp. 185–222). Edward Elgar.

Scott, W. R., & Davis, G. F. (2015). Organization: overview. In *International Encyclopedia of the Social & Behavioral Sciences* (Vol. 16, 2nd ed., p. Elsevier). https://doi.org/10.1016/B978-0-08-097086-8.73112-6

Senge, P. M. (1999). *The fifth discipline. The Art & Practice of the Learning Organization*. Random House Business Books.

Seuring, S., & Gold, S. (2013). Sustainability management beyond corporate boundaries: From stakeholders to performance. *Journal of Cleaner Production, 56*, 1–6. https://doi.org/10.1016/j.jclepro.2012.11.033

Smith, C. (2003). The new corporate philanthropy. In *Harvard business review on corporate responsibility*. Harvard Business School Press.

Soyka, P. A. (2012). *Creating a sustainable organization*. FT Press.

Stacey, R. D. (1993). *Strategic management and organisational dynamics*. Pitman Publishing.

Steensma, H. K. (1996). Acquiring technological competencies through inter-organizational collaboration: An organizational learning perspective. *Journal of Engineering and Technology Management - JET-M, 12*(4), 267–286. https://doi.org/10.1016/0923-4748(95)00013-5

Temel, M., Lozano, R., & Barreiro-Gen, M. (2021). Analysing the governance factors for sustainability in Organisations and their inter-relations. *Frontiers in Sustainability, 2*(June). https://doi.org/10.3389/frsus.2021.684585

The Economist. (2005). *The good company. A survey of corporate social responsibility*. The Economist.

Troy, L. C., Hirunyawipada, T., & Paswan, A. K. (2008). Cross-functional integration and new product success: An empirical investigation of the findings. *Journal of Marketing, 72*(6), 132–146. https://doi.org/10.1509/jmkg.72.6.132

Tschopp, D. J. (2005). Corporate social responsibility: A comparison between the United States and the European Union. *Corporate Social Responsibility and Environmental Management, 12*, 55–59.

Tunstall, D. (2005). In R. Lozano-Ros (Ed.), *Re: About the communication pyramid*.

Vaaland, T. I. (2004). Improving project collaboration: Start with the conflicts. *International Journal of Project Management, 22*(6), 447–454. https://doi.org/10.1016/j.ijproman.2003.11.003

Waddock, S., & Bodwell, C. (2007). *Total responsibility management*. Greenleaf.

Walsh, J. P., & Maloney, N. G. (2007). Collaboration structure, communication media, and problems in scientific work teams. *Journal of Computer-Mediated Communication, 12*(2), 712–732. https://doi.org/10.1111/j.1083-6101.2007.00346.x

Witjes, S., & Lozano, R. (2016). Towards a more circular economy: Proposing a framework linking sustainable public procurement and sustainable business models. *Resources, Conservation and Recycling, 112*, 37–44. https://doi.org/10.1016/j.resconrec.2016.04.015

Zafar, A., & Kantola, J. (2019). The effect of macro environmental and firm-level factors on open innovations in product development and their impact on firms' performance. In *International Conference on Applied Human Factors and Ergonomics* (Vol. 961, pp. 61–71). Springer. https://doi.org/10.1007/978-3-030-20154-8_6

Chapter 4
Tools, Initiatives, and Approaches Used for Implementing Sustainability

4.1 Introduction

From the 1970s, a number of tools, initiatives, and approaches (TIAs) have been developed to foster and implement sustainability (Johnson et al., 2018; Lozano, 2012; Robèrt, 2000). These have evolved from purely 'end-of-pipe' solutions (which are usually costly and inefficient) (Porter et al., 1995; Sarkis & Cordeiro, 2001) towards whole-system approaches, by changing products, processes, services, and systems (Holliday et al., 2002), so that waste is minimised, and resources used more efficiently and effectively in almost closed loops (McIntosh et al., 1998). It should be noted that the majority of the TIAs have been used in the corporate context (Lozano, 2012, 2020; Robèrt et al., 2002).

Twenty-four TIAs have been widely used and researched (Lozano, 2012, 2020). These can be divided according to their focus on operations and production, management and strategy, supply chain, and assessment and reporting.

4.2 Operations and Production-Oriented TIAs

The TIAs in this group have, by definition, a focus on the operations and production system element (see Chap. 3). This group includes (in alphabetic order) circular economy (CE), cleaner production (CP), design for the environment (DfE) eco-efficiency, Factor X, green chemistry, industrial ecology, and life cycle assessment (LCA).

Most of the TIAs in this group have a contribution to make for the environmental dimension of sustainability, e.g. through minimising waste and risk and achieving environmental improvements of products and processes. Some TIAs make a direct contribution to the economic dimension (e.g. CE).

R. Lozano, *Toward Sustainable Organisations*, Strategies for Sustainability,
https://doi.org/10.1007/978-3-030-99676-5_4

4.2.1 Circular Economy

Circular economy (CE) has been proposed as a concept to help address sustainability issues, particularly focusing on environmental issues (European Commission, 2018; Murray et al., 2015). CE was first proposed by Leontief (1928, 1991) in 1928, as a way of understanding causal relationships in the economic sphere through a combination of three ratios, called technical coefficients (cost, productivity, and distribution), to provide the best solution to achieve a circular flow. More recently, CE has focused, mainly, on environmental issues, as attested by the German Parliament law of 1996 on CE (Kreislaufwirtschaft) (Bilitewski, 2012), and the CE promotion in China and other Asian countries since the end of the 1990s (Andersen, 2006; Yong, 2007).

CE is based on "closing loops" through different types and levels of recovery (European Commission, 2014b; Preston, 2012; Urbinati et al., 2017; Yong, 2007; Yuan et al., 2008), from the 3Rs (reduce, reuse, and recycle) to the 9Rs (refuse, rethink, reduce, reuse, repair, refurbish, remanufacture, repurpose, recycle, recover) (Fernández & Kekäle, 2005; King et al., 2006; van Buren et al., 2016). This is achieved through the transformation of materials into useful goods and services based on resource efficiency (Klettner et al., 2014; Webster, 2013).

Three of the main definitions of CE (Pesce et al., 2020) are based on the (1) 3Rs for resources in production and consumption as à base for economic and social development resource recovery in production, circulation, and consumption (Liu et al., 2017); (2) Ellen MacArthur Foundation (2015), which states that "a circular economy is one that is restorative and regenerative by design and aims to keep products, components, and materials at their highest utility and value at all times, distinguishing between technical and biological cycles. This new economic model seeks to ultimately decouple global economic development from finite resource consumption"; and (3) the European Commision (2020), which defines CE as "a system where the value of products and materials is maintained for as long as possible through minimising waste and resource use, and reusing products when they reach their end of life, which can bring economic benefits, and contribute to innovation, growth, and job creation". The latter one is the most complete definition, since it provides a better connection between environmental and economic issues.

CE needs to include changes in technology for cleaner production, better reuse and recycling of waste (Bilitewski, 2012), prices that reflect full costs (Webster, 2013), social and organisational changes (European Commission, 2014a, 2014b), as well as economic and legal tools to promote the circular economy (Bilitewski, 2012). Switching to a circular economy needs the involvement and commitment of many stakeholders (European Commission, 2014a, 2014b) and a complete reform of the whole system of human activity, including production processes and consumption activities (Bilitewski, 2012; Yuan et al., 2008).

Two CE application categorisations have been proposed: firstly, one with three levels of activity focuses (Yong, 2007; Yuan et al., 2008) (1) the macro level, focusing on regions, cities, municipalities, or provinces; (2) the meso level, focusing on eco-industrial networks, where the waste (material or energy) from one company becomes the raw material of another; and (3) the micro level, which focuses on improving the environmental performance of an organisation. Secondly, the scope of CE activities, including assessment, review, individual, organisation, process, sector, cluster, and country/region (Lozano et al., 2021).

Most of the CE efforts have been at the macro and meso levels, with limited focus on the micro level, and mainly on companies (e.g. reducing food waste, eliminating hazardous waste, and recycling critical raw materials) (Barreiro-Gen & Lozano, 2020; Carpenter et al., 2018), with even more limited efforts on organisations, showing that there is a disconnect between the concept of CE and its application in practice (Barreiro-Gen & Lozano, 2020).

CE has focused on areas including food waste, hazardous waste, plastic waste, recycling of critical raw materials, illegal waste shipments, and phosphorus recycling (European Commission, 2014a, 2014b).

4.2.2 Cleaner Production (CP)

Cleaner production (CP) was developed by UNEP in Paris in 1989 (UNEP, 1990). CP is defined as the continuous use of integrated preventive strategies to process products and services, utilising raw materials, e.g. energy and water, efficiently to reduce waste at source and minimising risks to the environment and society (DeSimone & Popoff, 2000; UNEP, 2001). CP is aimed at achieving zero waste (i.e. the absolute reduction of the total use of all biomass and minerals on earth, Pauli, 1997), through environmental and economic approaches that halt emissions in production systems (Baumgartner & Zielowski, 2007; J. Ehrenfeld & Gertler, 1997; Kuehr, 2007). This is achieved by focusing particularly on the reduction and elimination of harmful emissions through technological solutions (Kuehr, 2007). In general, CP focuses on achieving environmental improvement in processes and product development (Glavič & Lukman, 2007; Pauli, 1997).

There are three levels of CP application, for (1) production processes, one or a combination of conserving raw materials, water and energy, eliminating toxic and dangerous raw materials, and reducing the quantity and toxicity of all emissions and wastes at source during the production process; (2) products, which aims to reduce the environmental, health, and safety impacts of products over their entire life cycles, from raw material extraction, through manufacturing and use, to the "ultimate" disposal of the product; and (3) services, which implies incorporating environmental concerns into designing and delivering services (UNEP, 2001).

CP should focus on changes in raw materials, processes, and products, as well as changing corporate culture and attitudes, applying know-how, and overcoming non-technical barriers (DeSimone & Popoff, 2000).

4.2.3 *Design for the Environment (DfE)*

Design for the environment (DfE), also called "eco-design", refers to the inclusion of environmental factors and considerations (such as material elimination or substitution, process optimisation, energy reduction, and product reuse) (DeMendonça & Baxter, 2001) in the design of the product or service (Holliday et al., 2002), which is a critical time for addressing environmental and economic impacts of a product (Paul T Anastas & Breen, 1997; Ashley, 1993). DfE encompasses eco-efficiency, health and safety, recycling, source reduction, and waste minimisation (Glavič & Lukman, 2007), which increases the potential to offer cost-effective take-back of products for recycling (Paul T Anastas & Breen, 1997; Ashley, 1993; DeMendonça & Baxter, 2001). DfE was developed mainly as a response to increased consumers' environmental awareness and tougher competition in the market respecting the environmental impacts of products (Hallstedt, 2008).

The European Commission established, in 2009, eco-design requirements for all energy using products (EC, 2009). There are more than 150 tools for eco-design applied to product design (Le Pochat et al., 2007), and some of the most typical are presented in Fig. 4.1. It should be noted that some of these tools have failed since they are not focused on new products, while working from the retrospective analysis of existing products (Borchardt et al., 2011). An example of DfE is IBM's Environmentally Conscious Products programme that has resulted in modifications to product development processes, where environmental attributes are routinely considered in decisions about new products (Ashley, 1993).

In spite of the apparent benefits of eco-design, such as competitiveness, cost reduction, new markets, and new products launches, the literature describes more pilot cases of eco-design than cases of its effective implementation during the product development process (J. Ehrenfeld, 2004).

Eco-Efficiency

Eco-efficiency started as an initiative of the World Business Council for Sustainable Development (see WBCSD, 2000), while the latter is of the UNEP (see UNEP, 2001; WBCSD & UNEP, 1998). The term eco-efficiency is a contraction of ecological and economic efficiency (Willard, 2002a). It is defined as a ratio of some economic value added (e.g. quantity, net sales, profits, or income) in relation to some measure of environmental impact (e.g. energy consumption, material consumption, water consumption, or greenhouse gas emissions) (J. R. Ehrenfeld, 2005); see Eq. 4.1. Eco-efficiency's aim is to link environmental and business excellence, i.e. making profits by using fewer natural resources, with less waste and emissions within the earth's carrying capacity (DeSimone & Popoff, 2000; Ekins, 2005; Hamann, 2003). It is one of the concepts most widely accepted as the business link to sustainability (Dyllick & Hockerts, 2002; Korhonen, 2003).

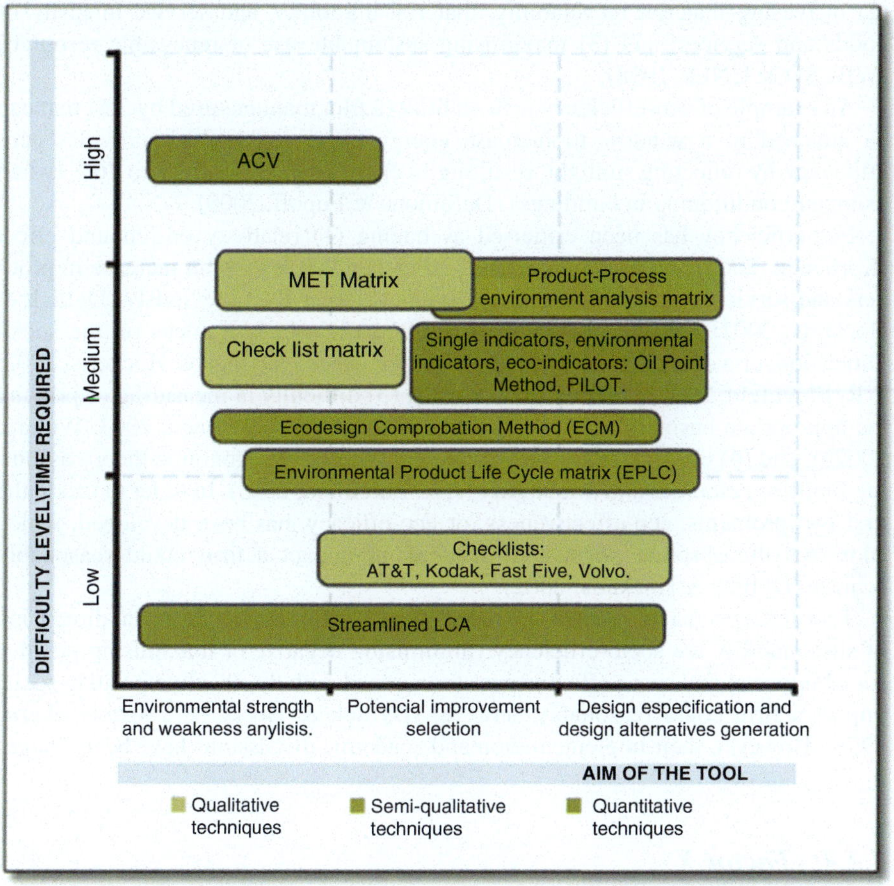

Fig. 4.1 Classification of methods for evaluating the environmental requirement of a product. Source: (Bovea & Pérez-Belis, 2012)

$$Eco-\text{efficiency} = \frac{\text{product or service value}}{\text{environmental influence}} \tag{4.1}$$

Eco-efficiency is guided by four principles (Holliday et al., 2002): (1) dematerialisation, by substituting knowledge flows for material flows or through customisation; (2) production loop closure, by working continuously towards closed-loop production systems and zero-waste factories wherein every output is returned to natural systems as a nutrient or becomes an input in the manufacture of another product; (3) service extension, by offering products that have extended durability or products that can be leased; and (4) functional extension, by designing and manufacturing products with newer and enhanced functionalities and services.

Eco-efficiency is aimed at (1) reducing the material intensity of goods and services, the energy intensity of goods and services, and toxic dispersion;

(2) increasing material recyclability, material durability, and service intensity of goods and services; and (3) maximising sustainable use of renewable resources (WBCSD & UNEP, 1998).

An example of eco-efficiency is Scotchtint (a film manufactured by 3 M that can be attached to a window to increase energy efficiency), which provides eco-efficiency by reflecting sunlight, resulting in energy savings of 55% to 76% (when using air conditioning in buildings) (DeSimone & Popoff, 2000).

Eco-efficiency has been criticised as having (1) relativity or rebound effect (Korhonen, 2003), i.e. improvements in efficiency that lead to an increase in products and services, making the absolute impacts larger than previously (Dyllick & Hockerts, 2002; Jansen, 2003; Jenkins, 2002); (2) a lack of focus on the social dimension, or a confusion with health and safety issues (Dyllick & Hockerts, 2002; J. R. Ehrenfeld, 2005; Holliday et al., 2002); (3) difficulty in measuring or proving the appropriate business case to quantify opportunities (Korhonen, 2003; Willard, 2002b); and (4) eco-efficiency being affected by standard economic theory assuming limitless resources and availability (J. R. Ehrenfeld, 2005). In order to tackle the first two problems, eco-effectiveness, or eco-efficacy, has been developed, which addresses the absolute positive environmental impact a firm could reasonably achieve (Dyllick & Hockerts, 2002).

Two concepts that are aimed at expanding eco-efficiency to the social dimension of sustainability are socio-efficiency (minimising negative or maximising positive social impacts) and socio-effectiveness (concerned with the absolute positive social impact a firm could reasonably achieve) (Dyllick & Hockerts, 2002; Hockerts, 1999). However, their implementation and academic discussions have been scarce.

4.2.4 Factor X

Factor X refers to the eco-efficiency initiatives Factor 4, Factor 5, and Factor 10, developed by the Wuppertal Institute (Robèrt, 2000; UNU, 2007; von Weizsäcker et al., 1998). They are based on reductions in turnover of resources on a global scale (Robèrt, 2000), i.e. increasing by a factor of "x" the amount of wealth that is extracted from one unit of a natural resource (DeSimone & Popoff, 2000; Holliday et al., 2002; von Weizsäcker et al., 2009). Factor X is qualitatively similar to eco-efficiency, but its focus is more quantitative (Reijnders, 1997).

Factor X has been received with mixed reactions; while in the USA its use has been restricted to environmental NGOs, in Europe (mainly in Austria, Germany, and the Netherlands), it has had a positive impact on environmental policies. Nonetheless, there has been no substantial technological or economical enforcement for its implementation in such countries (Reijnders, 1997).

Factor 4 principles were applied by a team in the Rocky Mountain Institute in the design for a car with 57% lighter structure and 34% lighter chassis components, allowing a 38% lighter propulsion system. Including reductions in the mass of the electrical system, trim, and fluids, the overall vehicle was 52% lighter (von Weizsäcker et al., 2009).

4.2.5 *Green Chemistry (GC)*

GC focuses on the use of chemical techniques to reduce or eliminate the use, or generation, of feed-stocks, products, by-products, solvents, reagents, or other hazardous chemicals that are, or might be, dangerous to human health or the environment (Paul T Anastas & Breen, 1997). GC is aimed at preventing waste before it is ever formed by considering the environmental impact, or potential impact, of a product or process (Paul T. Anastas & Warner, 1998). GC relies on 12 rules based on 5 principles (waste minimisation, renewable resources, eco-efficiency, degradation, and health and safety) that are aimed at designing or modifying chemical reactions to be more environmentally friendly (Glavič & Lukman, 2007). It has been estimated that chemical production using green chemistry accounts for only 1 per cent of chemical industry production as a whole (Matus, 2010; Matus et al., 2012).

The key points in the Plan of Implementation from the World Summit on Sustainable Development (UN, 2002) regarding chemicals were that (1) they are used and produced in ways that minimise the adverse effects on human health and the environment; (2) developing countries are supported in strengthening their capacity for the sound management of chemicals and hazardous wastes; and (3) activities aimed at fostering sound management of chemicals and hazardous wastes are promoted.

The three approaches to GC are business models, policy initiatives, and science. **Business model** initiatives are based on a move away from the supplier seeking to increase sales to increase profits and the buyer seeking to decrease the volumes purchased to reduce costs, where the key principle in both concepts is to align incentives so that the supplier becomes the service provider while the buyer becomes the service user. These include the Strategic Approach for International Chemicals Management (SAICM) policy framework, adopted at the International Conference on Chemicals Management held in Dubai in 2003 (United Nations Environment Programme & UNEP, 2006), and the Global Chemical Leasing (ChL) programme, launched by the United Nations Industrial Development Organization (UNIDO) (Jakl & Schwager, 2008; Lozano et al., 2014; Lozano, Carpenter, et al., 2013; UNIDO, 2011).

ChL approaches have resulted in 100% recovery of nickel and iron in electroplating (Schwager & Moser, 2005), over 98% recovery of platinum for future fuel-cell vehicles (Kromer et al., 2009), and 75–80% of a vehicle is recycled or reused, following the European Union end-of-life directive (Gerrard & Kandlikar, 2007). Some of the key benefits for the ChL partners include (Lozano, Carpenter, et al., 2013) increased ability to create value for both parties, flexibility and speed of joint responses to changing market or customer needs and expectations, and optimisation of costs and resources. This can lead to establishing and improving partner relationships that balance short-term gains with long-term considerations between partners, such as communicating more clearly and openly, sharing information and future plans, and establishing joint development and improvement activities.

Policy initiatives demand the introduction of safer alternative chemicals, including the Registration, Evaluation, Authorisation and Restriction of Chemicals (REACH) (European Commission & European Commission – Environment, 2012), control of Waste Electrical and Electronic Equipment (WEEE) (EU, 2003), Responsible Care (a chemical industry's global voluntary code of conduct under which companies, through their national associations, work together to continuously improve their health, safety, and environmental performance) (ICCA, 2011; Prakash, 2000), and adherence to Chemical Industry Association principles (P. T. Anastas, 2000).

Science initiatives are aimed at providing the scientific solutions necessary to achieve those policy initiatives. An example of this is BASF's SavantTM, made from nylon six fibres, which can be depolymerised and reused. BASF has initiated a take-back programme for used nylon carpet, rather than down-cycling it into a material of lesser value. Thus the used nylon is "upcycled" into a product of high quality (Heine, 2007).

4.2.6 Industrial Ecology (IE)

Industrial ecology (IE) refers to the restructuring of industry in the form of an ecosystem with materials (including raw materials and wastes) flowing through interconnections of production processes (J. Ehrenfeld, 2004; Isenmann, 2003; Lowenthal & Kastenberg, 1998). IE is a vision of an interconnected industrial system in which new products evolve out of, or consume, available waste streams and where processes are in turn developed to produce usable "waste" (DeSimone & Popoff, 2000). This is achieved by optimising industrial systems through a whole-system perspective (EC, 2009; Parker, 2011), with a closed-loop model resembling that of ecosystems (Lehman, 1999).

The objective of IE is to treat materials and energy, considered as by-products or waste, as raw materials by other companies (DeSimone & Popoff, 2000; Heeres et al., 2004; Lowe & Evans, 1995). This is done on the firm level, with design for environment, pollution prevention, eco-efficiency, and green accounting; across firms, with industrial symbiosis, life cycle analysis, and industrial sector initiatives (Gibbs & Deutz, 2007); and at the regional/global level, including budgets, materials and energy flow studies, dematerialisation, and decarbonisation (Jacobsen, 2006; Warhurst, 2002).

IE has, as inherent benefits, the reduction of pollution, a decrease in natural resource usage, and, it is claimed, the creation of new jobs (Lehman, 1999). Its main challenges are providing sufficient and correct exchange of information about the different industries and their inputs and outputs (EC, 2009) and achieving integration across different fields and overcoming of barriers, such as cultural (EC, 2009), technical, economic, information, organisational, and regulatory/legal (EC, 2009; Warhurst, 2002).

The classical example of IE is the seaside industrial town of Kalundborg, Denmark, where 11 physical linkages comprise much of the tangible aspects of IE (EC, 2009), with estimated savings of between US$ 12 and 15 million per year (Warhurst, 2002). The Kalundborg case was not designed as IE, but it evolved gradually over a number of decades (Warhurst, 2002).

4.2.7 Life Cycle Assessment (LCA) and Life Cycle Management (LCM)

The first life cycle studies took place in the 1960s and 1970s between different materials and packaging options. Their purpose was to evaluate alternatives from the point of view of, e.g. energy consumption or waste management. The first life cycle studies conducted by companies were frequently related to attempts to defend themselves against environmental requirements by demonstrating that the problem is more complex than was initially believed (Heiskanen, 2002).

LCA is a technique aimed at evaluating all processes in the life cycle of a product or service, from downstream (i.e. extraction) to upstream (i.e. disposal), including use (DeSimone & Popoff, 2000; Holliday et al., 2002; Robèrt, 2000). It focuses, primarily, on quantifiable information that can help in the decision-making process (Hale, 1996). It is based on natural scientific principles, such as the laws of thermodynamics, and on the collection of data from observable processes and is often presented as an objective, scientific method, or at least as a method that strives towards these goals (Heiskanen, 2002).

LCA is used for two main purposes, to (1) assess the environmental impacts throughout the life cycle stages in order to detect "hotspots" of a product or service (Finkbeiner et al., 2006; Robèrt et al., 2002) and (2) compare the environmental impact of products or services (Heiskanen, 2002). For the latter, a "functional unit" must be established, where environmental burdens in multi-product processes are allocated or where average or marginal energy emission data are used (Heiskanen, 2002).

LCAs are complex because they almost always involve uncertain consequences relative to multiple criteria (Seppälä et al., 2002). It is necessary to determine how the inputs and outputs should be weighted to reflect their environmental risks (Ashley, 1993). LCAs are time-consuming, expensive, and data-intensive (Hale, 1996). They can be cumbersome due to the data requirements and boundary setting for the analysis (Ashley, 1993). Although LCA is considered to be one of the most rigorous sustainability tools, it entails trade-offs between specificity and depth and comprehension and applicability (Holmberg & Robert, 2000).

A related approach is life cycle management (LCM), which is a business management process that can be used by all types of business (and other organisations) in order to improve their sustainability performance (UNEP & SETAC, 2007). LCM can be used equally by both large and small firms. Its purpose is to ensure more sustainable value chain management (Haes et al., 2002). It can be used to target, organise, analyse, and manage product-related information and activities in relation to continuous improvement over the product life cycle (UNEP & SETAC, 2007).

4.3 Management and Strategy-Oriented TIAs

The TIAs in this group have, by definition, a focus on the management and strategy system elements (see Chap. 3). This group includes (in alphabetic order) corporate citizenship (CC), corporate social responsibility (CSR), (more) sustainable business models (SBMs), sustainable/socially responsible investment (SRI), the natural step (TNS), and the triple bottom line (TBL).

The TIAs in this group have a better coverage of the economic (e.g. business models, customers, investment, and profits) and social (e.g. focusing on health and safety, stakeholder management, philanthropy, and societal welfare) dimensions of sustainability that those in the operations and production group.

4.3.1 Corporate Citizenship (CC)

Corporate citizenship (CC) is considered to be a core business strategy tool (Birch & Littlewood, 2004), where corporations have social rights and responsibilities to their stakeholders beyond wealth maximisation (Carroll, 1998; Leisinger, 2003; McIntosh et al., 1998). This includes compliance with laws and regulations, ethical behaviour, and contributions to social and economic welfare (Birch & Littlewood, 2004; Carroll, 1998).

CC is used as a synonym for concepts such as business ethics, corporate environmental management, business and society, business and governance (Hopkins, 2002; Matten & Moon, 2004; Swift & Zadek, 2002), and sometimes CSR (Frankental, 2001; Langer & Schön, 2003; Matten & Moon, 2004).

4.3.2 Corporate Social Responsibility (CSR)

CSR practices can be traced almost as far back as the French Revolution (Frankental, 2001); the origins of the "modern" form of CSR are subject to discussion. Academically, CSR began in the wake of the Great Depression, in the late 1920s (Dodd Jr., 1932; Lantos, 2001). Since then, many CSR definitions have appeared, from which the key points being addressed include stakeholder engagement and participation (C.E.C., 2001; Holme & Watts, 2000); product impact; health and safety; dealing with corruption (Holme & Watts, 2000); human rights and freedom of association (C.E.C., 2001; Holme & Watts, 2000; UNGC, 2008; Welford, 2005);

communication, reporting, disclosure, and transparency (Holme & Watts, 2000); and environmental protection and management of resources (C.E.C., 2001; Elkington, 2002; Holme & Watts, 2000). CSR can be defined as "the continuing commitment by the business to behave ethically and contribute to economic development while improving the quality of life in the workplace and of the workforce, as well as the local community and society at large" (Holme & Watts, 2000).

CSR aims at improving societal welfare and well-being (Farmer & Hogue, 1973; Frehs, 2003; Mintzberg, 1983). It integrates social and environmental concerns, and stakeholder interactions, into a business's operations (C.E.C., 2001; Frehs, 2003) and into the long-term prosperity of the corporation (Holme & Watts, 2000). CSR considers that corporations are responsible for their wider impacts on society (Frankental, 2001). CSR is voluntary (C.E.C., 2001) since, typically, it goes beyond legal expectations and compliance, investing more into human capital, the environment, and stakeholder relations (C.E.C., 2001, 2002; Frehs, 2003). However, it should not be a substitute for governmental regulations and legislation (Swift & Zadek, 2002; UNIDO et al., 2002). CSR should be about how businesses are managed, and not an optional "add-on" (C.E.C., 2002).

Some of the key CSR issues, among a long list, are monitoring child labour (MacLeod & Lewis, 2004; Zadek, 1999); stakeholder engagement and participation, including local communities, employees, shareholders, business partners, suppliers, customers, public authorities, and NGOs (C.E.C., 2001; Holme & Watts, 2000); innovation stimulation (Frehs, 2003); lifelong learning, equal opportunities, and better management of natural resources (C.E.C., 2001); communication, reporting, disclosure, and transparency (Holme & Watts, 2000); product impact (Holme & Watts, 2000); health and safety (MacLeod & Lewis, 2004); dealing with corruption (Holme & Watts, 2000); human rights, freedom of association, vocational education, fair wages, equal opportunities, non-discrimination (Holme & Watts, 2000; MacLeod & Lewis, 2004; UNGC, 2008; Welford, 2005), protection of local suppliers, child labour, labour standards, ethics, indigenous people, and fair trade (Welford, 2005); and environmental protection (Holme & Watts, 2000) that goes beyond pollution control (Elkington, 2002).

CSR has been driven mainly by large corporations, with some complementary efforts by SMEs and cooperatives (C.E.C., 2001, 2002). Internally, one of the main drivers in large corporations has been ethical leadership (Szekely & Knirsch, 2005). Other internal drivers include risk management and protection of business reputation (Lantos, 2001), improvements in economic values (C.E.C., 2001; Carroll, 1999; Lantos, 2001), and enhancement of corporate image (Frehs, 2003). National policies have played an important role in driving CSR, from outside the corporation, for example, the proactive measures of some European Union countries (MacLeod & Lewis, 2004), and social action under administrative guidance (*gyosei-shido*) in Japan (Fukukawa & Moon, 2004). Other external drivers have been NGOs and stakeholder pressure (Frehs, 2003; Zadek, 1999).

Table 4.1 presents the motivations to engage with CSR, divided into overarching motivations (those that have effects inside and outside the corporation), internal motivations (dealing with processes inside the corporation), and external motivations (relations with external stakeholders).

Table 4.1 Motivations to engage in CSR

Overarching motivations	
• Meet and exceed stakeholder expectations • Behave ethically • A belief that corporations must earn their "licence to operate" • Avoidance of fines and penalties	
Internal motivations	External motivations
• Attract and retain employees • Help improve trust within the company, i.e. stronger employee motivation and commitment • Have a more compliant workforce • Increase employee productivity • Help to increase product quality • Help boost innovation and innovative practices • Help manage risks, intangible assets, and internal processes • Improve performance and generate more profits and growth • Reduce costs, while improving process efficiencies and reducing waste	• Help improve trust outside the company, i.e. with business partners, suppliers, consumers, and others • Improve relations with regulators and ease access to permits • Improve access to markets and customers • Improve customer satisfaction • Help to restore trust in corporations • Help enhance corporate and brand reputation • Reduce or eliminate pressures from NGOs

Sources: adapted from (C.E.C., 2001, 2002; Frankental, 2001; Frehs, 2003; Fukukawa & Moon, 2004; Laffer et al., 2004; Lantos, 2001; The Economist, 2005)

Table 4.2 Differences in CSR perspectives in the USA and Europe

The USA	Europe
• Corporations considered as aggregated entities. • The corporation can be created de facto. • The corporations is seen as an autonomous "person". • The corporation is a fictitious entity without rights and responsibilities. • The only duty is towards its shareholders.	• Corporations are considered as real entities. • The corporation is *persona representata* (it acts through its agents). • The corporation, per se, can be accused of committing certain crimes. • The corporation has duties to stakeholders including the environment.

Sources: Synthetised from (Avi-Yonah 2005; Berle et al., 1947; Jensen & Meckling, 1976; Millon, 1990; Radin, 1932; Smith, 2003)

CSR is being interpreted differently in Europe and in the USA (as highlighted in Table 4.2). In Europe the mainstream corporate entity is more open and flexible towards CSR, encompassing, in general, environmental and social aspects (C.E.C., 2001, 2002; Porter & Kramer, 2003; Smith, 2003). In the USA CSR is usually a synonym for corporate philanthropy (Porter & Kramer, 2003; Smith, 2003).

In general, it is possible to observe an evolution from corporate philanthropy (see Porter & Kramer, 2003) to a more broad approach by treating stakeholder responsibilities, from a social perspective (see Hopkins, 2002; Reinhardt, 2000), and then to more recent concerns that aim to integrate environmental matters (Avi-Yonah 2005; C.E.C., 2001, 2002; Carroll, 1999; WBCSD, 2002).

For some authors, CSR increases costs and impairs performance (Farmer & Hogue, 1973; Henderson, 2005), attempts fundamental reform of capitalism in order to make it more humane (The Economist, 2005), distracts attention from genuine business ethical problems by taking into account stakeholders (Henderson, 2005; Mintzberg, 1983; The Economist, 2005), and is merely a cosmetic exercise for public relations purposes (Frankental, 2001; Mintzberg, 1983; The Economist, 2005).

For other authors CSR is limited since it is (1) difficult to demonstrate positive correlations between CSR and "the bottom line" (Avi-Yonah 2005; Laffer et al., 2004); (2) difficult to evaluate a corporation's performance against all the CSR issues (Avi-Yonah 2005); (3) considered a panacea to solve all the world's problems, such as global poverty, social exclusion, and environmental degradation (van Marrewijk & Hardjono, 2003); (4) difficult to operationalise CSR definitions and principles (Lozano, 2012); (5) only profitable companies can engage in CSR (Laffer et al., 2004); (6) by engaging with CSR, the corporation might take on roles that belong to the government, for example, building and running schools, hospitals, and roads (Ite, 2004); also (7) the environment is not explicit in the CSR term (Fukukawa & Moon, 2004); and (8) there are too many, often contradictory, CSR definitions (Lozano, 2012).

4.3.3 Corporate Sustainability (CS)

Corporate Sustainability (CS) emerged as an alternative to CSR (Lozano, 2012), where CS is considered to be a precondition for doing business, as a "business case" (Dyllick & Hockerts, 2002), and the desirable path for organisations (Dunphy, Griffiths, Benn, 2003; Weymes, 2004). It was first defined as "Meeting the needs of a firm's direct and indirect stakeholders (such as shareholders, employees, clients, pressure groups, communities etc.), without compromising its ability to meet the needs of future stakeholders as well" (Dyllick & Hockerts, 2002). However, a more complete definition is "Corporate activities that proactively seek to contribute to sustainability equilibria, including the economic, environmental, and social dimensions of today, as well as their inter-relations within and throughout the time dimension, whilst also addressing the company's system (including Operations and production, Management and strategy, Organisational systems, Procurement and marketing, Governance and Assessment and communication), and its stakeholders" (Lozano, 2012; Lozano et al., 2017).

Corporate Sustainability (CS) has been proposed as a framework to address the full array of sustainability challenges and issues (Lozano, 2012) (see Bartelmus, 1999; Dyllick & Hockerts, 2002). CS has to encompass a holistic perspective (Linnenluecke et al., 2009; Lozano, 2013b) and include the four dimensions of sustainability (economic, environmental, social, and time), as well as their interactions (Lozano, 2008). In order for a company to become more sustainability-orientated, it should engage in changes that include the introduction of resource-efficient technologies, sustainability reporting schemes, and providing sustainable products, services, and product-service combinations (Siebenhüner & Arnold, 2007).

4.3.4 Sustainable/Socially Responsible Investment (SRI)

Traditionally, there has been pressure on public companies from financial markets to maximise short-term results (Focusing capital on the long term, 2014). Sustainability is challenging such positions, especially when investors, with significant and long-term holdings (Gompers et al., 2003), are concerned about their client's actions (Aras & Crowther, 2008). Investors promote change within a sector through the allocation of capital due to raised environmental ratings, leading to an increase in return on assets (Russo & Fouts, 1997).

An increasing number of investors have begun to integrate environmental, social, and governance criteria into their investment decision-making process (Chava, 2010; Eurosif, 2014; Wai et al., 2017). Sustainable/socially responsible investment is aimed at integrating environmental, social, and governance criteria into the investment decision-making process (Chava, 2010; Eurosif, 2014; Wai et al., 2017). The integration includes social performance (e.g. employment quality, safety and health, training, and development), environmental performance (including emissions reductions, resource reduction, and product innovation), economic performance (such as client loyalty, performance, and shareholder loyalty), and corporate governance performance (including board structure, compensation policy, and vision and strategy) (Escrig-olmedo et al., 2017). By integrating the long-term perspective, investors are increasingly addressing the four dimensions of sustainability—economic, environmental, social, and time (Lozano & Reid, 2017). In the context of ESG, the integration includes social performance (e.g. employment quality, safety and health, training and development), environmental performance (including emissions reductions, resource reduction, and product innovation), economic performance (such as client loyalty, performance, and shareholder loyalty), and corporate governance performance (including board structure, compensation policy, and vision and strategy) (Escrig-olmedo et al., 2017).

4.3.5 (More) Sustainable Business Models

Sustainability has challenged traditional business models (Lozano, 2012; Murray, Skene, & Haynes, 2015), which has resulted in a shift from selling products to providing service solutions to meet customer needs (Lay, Schroeter, & Biege, 2009; Mont, Dalhammar, & Jacobsson, 2006) and better engagement with stakeholders, while creating competitive advantages to customers, the company, and society (Boons & Lüdeke-Freund, 2013; Porter & Kramer, 2011; Stubbs & Cocklin, 2008).

A number of sustainable business models (SBMs) have been proposed to better contribute to sustainability (as proposed by Bohnsack, Pinkse, & Kolk, 2014; Mont et al., 2006; Schaltegger, Hansen, & Lüdeke-Freund, 2016; Stubbs & Cocklin, 2008), such as (1) leasing, where the supplier becomes a service provider by retaining the ownership and assuming responsibility for maintenance; in this case the customer pays a regular fee for unlimited individual access to the product; (2) renting, similar to leasing, however, the customer does not have unlimited access to the product; and (3) "product pooling", where the equipment is used simultaneously by several users instead of a sequential mode of use (Lay et al., 2009). Three examples of sustainability oriented business models are selling the function that baby prams provide (Mont et al., 2006); chemical leasing (ChL), where chemical companies move from selling tons of chemicals to a service oriented business (Lozano et al., 2014); and ride- and car-sharing business models for sustainability (Cohen & Kietzmann, 2014), such as carpooling, flexible carpooling, vanpooling, and ride-sharing.

The majority of SBMs proposals are focused on the environmental dimension, through ecological modernisation or through resource efficiency (see Bocken et al., 2014; DeSimone & Popoff, 2000; Holliday et al., 2002; Robèrt, 2000; Stubbs & Cocklin, 2008). A limited number focus on the social dimension of sustainability (e.g. Roome & Louche, 2016) focus on the social dimension. Only two proposals address explicitly the time dimension of sustainability (Lozano, 2018; Upward & Jones, 2016).

Most definitions tend to be based on a narrow, business-oriented perspective of value proposition, creation, and delivery (see Abdelkafi & Tauscher, 2015; Bocken et al., 2014; Boons & Lüdeke-Freund, 2013; Schaltegger et al., 2016), with the exception of Upward and Jones (2016). This arises from the traditional BM perspective, i.e. from a (mainly economic) value point of view; however, a SBM should be seen from a sustainability perspective on how to add value to the four dimensions of sustainability, and not from value focused on how to increase sustainability performance (Lozano, 2018).

From the above discussions, it can be highlighted that the elements in a business model are comprised of:

- Material resources and energy. The inputs are transformed in the company through operations and production (using technology) into output products and/or services. These inputs and outputs are part of the supply chain (see Bocken et al., 2014; Lozano, 2012).
- Stakeholders that affect and are affected by the company (see Boons & Lüdeke-Freund, 2013; Lozano, 2012; Stubbs & Cocklin, 2008).
- Economic value (intrinsic to the material inputs and explicit to the output products/services) and value added in the company through resource efficiency and transformation (see Bocken et al., 2014; Boons & Lüdeke-Freund, 2013; Stubbs & Cocklin, 2008).
- Human resources that enter and leave the company (see Penrose, 1959).
- Tangible resources (e.g. infrastructure, plant equipment, land and natural resources, waste products, and finished goods) (see Penrose, 1959).
- The (intrinsic) environmental value of the materials and energy and of the output products/services (implied by Bocken et al., 2014; Boons & Lüdeke-Freund, 2013; Stubbs & Cocklin, 2008).
- Waste, some of which may be recycled, but due to the laws of conservation and mass, and energy, some waste will exit the system (Lozano, 2018).

The SBM definition that encompasses all the previous elements states: "A holistic and systemic reflection of how a company operationalises its strategy, based on resource efficiency (through operations and production, management and strategy, organisational systems, governance, assessment and reporting, and change), so that the outputs have added value and contribute to sustainability more than the inputs (in regards to material and resources that are transformed into products and services, economic value, human resources, and environmental value). The business model is affected by the company's resources (tangible and intangible), the supply chain, and the company's stakeholders (internal, inter-connecting, and external), including the environment (inside and outside the company)" (Lozano, 2018).

Most authors use the term "sustainable business model", with the exception of Schaltegger et al. (2016), who prefer the term "business models for sustainability". A better term would be "more sustainable business models", where sustainability is embedded in the business model and is based on sustainability as a dynamic ideal, and thus no business model will ever be fully sustainable (Lozano, 2018).

4.3.6 The Natural Step (TNS)

The natural step (TNS) is an international educational organisation dedicated to accelerating society's movement towards SD (Robèrt et al., 2002; Willard, 2002b), with a framework to aid in this transition (Robèrt, 2000). The framework aims to redirect people's attention away from detailed environmental problems far "downstream" in cause-effect chains (e.g. addressing every new chemical with a new risk assessment), instead of focusing upstream on the unifying causes behind many problems (such as the general increase in the systematic concentration of man-made chemicals in nature) (Doppelt, 2003). TNS is built on back-casting, i.e. envisioning a desirable future and working to move to that point (Robèrt et al., 2002).

4.3.7 The Triple Bottom Line (TBL)

The TBL was first proposed by Elkington (1998). It focuses on incorporating environmental and social performance indicators, while complementing and balancing the economic indicators, into company management, measurement, and reporting processes (Atkinson, 2000; Elkington, 2002; Frankental, 2001; Wilenius, 2005), as illustrated in Fig. 4.2. TBL aims to question a company's values, strategies and practices, and how these can be used to achieve SD (Milne et al., 2003). The TBL has become, for some companies, the concept representing their actions and contribution to SD (Milne et al., 2003; Verdeyen et al., 2004).

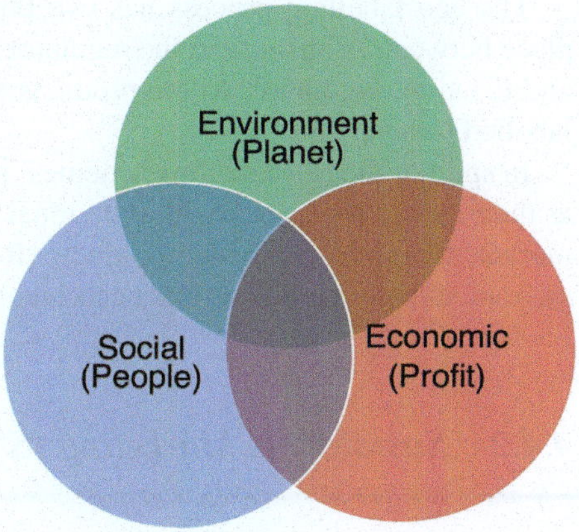

Fig. 4.2 The triple bottom line. Source: adapted from (Elkington, 1998)

4.4 Supply Chain-Oriented TIAs

The TIAs in this group have, by definition, a focus on the supply chain system element (see Chap. 3). This group includes (in alphabetic order): eco-labelling, sustainable marketing (SM), sustainable livelihoods (SL), and sustainable supply Chains (SSC).

The TIAs in this group tend to have a balanced coverage of the economic, environmental, and social dimensions, with the exception of eco-labelling which is limited in its coverage of social issues.

4.4.1 Eco-Labelling

Eco-labelling is based on a market approach to the protection of the environment (Hale, 1996; OECD, 1997). It aims to inform consumers of the environmental impacts throughout the production, consumption, and waste phases of products and services and, to a great extent, influence consumers' behaviour towards more environmentally friendly consumption patterns (Hale, 1996; Nadaı, 1999; Rex & Baumann, 2007). It also aims to encourage producers, governments, and other agents to increase the environmental standards of products and services (Galarraga Gallastegui, 2002).

Eco-labelling has increasingly become popular among corporations and policymakers. During the last two decades, several national eco-labelling programmes have appeared (Nadaı, 1999; OECD, 1997; Rex & Baumann, 2007). Some examples of eco-labels include the German Blue Angel, the Nordic Swan, the Swedish Environmental Choice, the EU Eco-Label Award Scheme, the Canadian Environmental Choice Program, the Green Seal operated in the USA, the Japanese Eco-Mark, and the French NF Environnement (OECD, 1997).

The eco-labelling process has two phases: (1) the negotiation phase that takes place between the firms and the regulator in order to set up the eco-labelling criteria and (2) a market phase, where products that meet the criteria can be appropriately labelled (Nadaı, 1999).

In spite of the eco-labelling schemes' promises, they have some drawbacks, such as their moderate success with individual consumers (OECD, 1997), the low overall market share of eco-labelled products (Rex & Baumann, 2007), and their lack of a guarantee that effective product eco-labels have been developed (Nadaı, 1999).

4.4.2 Sustainable Marketing (SM)

Sustainable marketing (SM) focuses on promoting re-consumption, influencing consumer choice behaviour through communications, aligning the marketing mix of existing products with sustainability issues, and promoting appropriate changes in corporate structure and engaging all stakeholders so as to facilitate the long-term use of these strategies (Peattie, 2001; Sheth et al., 1995).

Sometimes, green marketing is used as a synonym for sustainable marketing, but they are different. Green marketing integrates environmental considerations into marketing activities, i.e. it reduces the environmental impacts of both products and production systems (McDaniel & Rylander, 1993; Peattie, 2001). It also aims to identify and satisfy the requirements of green consumers (Peattie, 1995). This has three implications; it should (1) cater for the needs of the people who consume the product and should also consider the needs of those who do not, (2) develop products based on the requirements and expectations of both present and future customers and non-customers, and (3) address environmental issues, e.g. biodiversity loss, climate change, and ozone depletion (Peattie, 2001).

A related concept is social marketing, which uses marketing as a way to influence a target audience to focus on social issues, i.e. accept, modify, or abandon a behaviour for the benefit of individuals, groups, or society as a whole (Kotler & Armstrong, 2001).

4.4.3 Sustainable Livelihoods (SL)

Sustainable livelihoods (SL) focuses on creating new businesses and markets that benefit both the poor and the company. As opposed to philanthropy, SL aims to align the company with the laws, norms, expectations, and aspirations of the society in which it operates. It has a conscious focus on the poor as aspiring customers and business partners (WBCSD, 2004). SL aims to benefit society by focusing on, in addition to employment, natural resource management, redistribution of livelihood resources, prices and payments, and health, while abolishing restrictions and hassle, while providing safety nets for poor people during bad times (Chambers, 1995). SL is based on providing the skills and assets necessary for people to live reasonably secure lives, to cope with and recover from stress and shocks, and to provide opportunities for the next generation (Chambers, 1995; WBCSD, 2004). Usually, SL reflects the combination of a focus on core competencies that facilitate innovation, partnerships across sectors (e.g. governments and NGOs), and the localisation of value creation (WBCSD, 2004).

An example of SL is the South African Sugar Association's (SASA) joint venture with provincial departments of agriculture, aimed at building an effective and well-coordinated extension service for small-scale farmers through mentorship in business skills, workshops on cane growing, grassroots leadership, and business and functional training sessions (WBCSD, 2004).

4.4.4 Sustainable Supply Chains (SSC)

Sustainable supply chains (SSC) deal with the planning, execution, and control of integrating economic, environmental, and social issues to improve the long-term performance of an individual company and its supply chain (Stindt, 2017). This involves the management of materials, information, capital flows, and cooperation among companies (Seuring and Müller 2008a, b). SSC promotes efficiency and synergies between business partners and lead corporations and helps enhance environmental and social performance, minimise waste, and achieve cost savings (Rao & Holt, 2005). Ahi and Searcy (2013) defined sustainable supply chains as "The creation of coordinated supply chains through the voluntary integration of economic, environmental, and social considerations with key inter-organisational business systems designed to efficiently and effectively manage the material, information, and capital flows associated with the procurement, production, and distribution of products or services in order to meet stakeholder requirements and improve the profitability, competitiveness, and resilience of the organisation over the short- and long-term".

In a SSC, the focal company initiates sustainability activities, forces upstream companies to engage with sustainability initiatives, and is decisive in designing the final product (Rao & Holt, 2005; Stindt, 2017).

Two types of relationships can take place in the SSC (Rao & Holt, 2005): (1) competition, which ensures compliance, increases profit margins, creates new market opportunities, and may strengthened business activities but threatens reputation, and (2) collaboration, which can lead to better efficiency and secure long-term benefits for all (DeSimone & Popoff, 2000). For more details about organisations' interactions, refer to Chap. 3.

Some of the challenges to implementing SSC management include insufficient dedication to sustainable development, lack of top management commitment and resources, emphasis on efficiency and governance of supply relationships rather than environmental and social sustainability, a reactive and firefighting culture, high costs, complexity of and insufficient communication in supply chains, green washing, and poor supplier commitment (Preuss, 2009; Seuring & Müller, 2008a, 2008b).

4.5 Assessment and Reporting-Oriented TIAs

The TIAs in this group have, by definition, a focus on the assessment and reporting system elements (see Chap. 3). This group includes (in alphabetic order): environmental and social accounting (ESA), environmental management systems (EMS), integrated management systems (IMS), and sustainability reporting (SR).

Some of the TIAs in this group have a clear focus on environmental issues (ESA and EMS), some on social issues (AA1000 and SA8000), while others concentrate on economic, environmental, and social issues (IMS and GRI).

4.5.1 Environmental and Social Accounting (ESA)

Environmental and Social Accounting's (ESA) objective is to ascribe monetary values to the direct and indirect environmental and social impacts of a company's activities using a variety of valuation methods (Burritt et al., 2009; de Beer & Friend, 2006; Warhurst, 2002). This has the aim of demonstrating the potential for environmentally beneficial investments to yield significant financial pay-offs, through the avoidance of environmental and social liabilities (de Beer & Friend, 2006). Some of the issues that are monetarised include air and climate; waste, including wastewater, soil and groundwater; noise and vibration; biodiversity and landscape; and radiation (de Beer & Friend, 2006), as well as costs of plant closure (Lehman, 1999). Although ESA can help assist companies in managing, measuring, and improving the environmental aspects of their operations (de Beer & Friend, 2006), its bases limit reform and focus on the corporation as the accounting entity, thus perpetuating the status quo (Lehman, 1999).

4.5.2 Environmental Management Systems (EMS)

Environmental Management Systems (EMS) are administrative tools aimed at assessing the environmental impact of the operations of organisations, mainly corporations, and for improving their environmental performance (Brorson & Larsson, 1999; Robèrt, 2000).

EMS are designed to help manage resources and improve the reliability and credibility of the environmental policy (Brorson & Larsson, 1999) and serve as environmental credentials for customers (Robèrt et al., 2002).

There are five common elements in all EMS: (1) identifying company impacts on the environment, (2) understanding current and future legal obligations, (3) developing plans for improvement, (4) assigning responsibility for implementation of plans, and (5) periodic monitoring of performance (DeSimone & Popoff, 2000).

Some of the advantages of EMS include providing a systematic understanding of environmental issues; reporting internally about results, performance, and plans; and worldwide recognition, whereas some of their disadvantages are that they are, sometimes, merely informational, and they are costly (Lozano & Huisingh, 2011).

Two of the most recognised EMS are the International Organisation for Standardization (ISO) 14,000 series and the EU Eco-Management and Audit Scheme (EMAS) (Brorson & Larsson, 1999; Robèrt, 2000).

ISO 14000 series is aimed at assessing the environmental impact of operations and improving their performance (Brorson & Larsson, 1999; ISO, 2021; Robèrt et al., 2002). ISO 14000 series reports on performance and plans internally (Morhardt et al., 2002). It is based on five main elements: (1) Identify impacts to the environment, (2) understand current and future legal obligations, (3) develop plans for improvement, (4) assign responsibility for plans implementation, and (5) periodic performance monitoring (DeSimone & Popoff, 2000).

The ISO 14000 series is comprised of (ISO, 2021):

- 14,001: Requirements, with guidance for use, often seen as the cornerstone of the series, and the only one that can be externally certified
- 14,004: General guidelines on principles, systems, and support techniques
- 14,015: Environmental assessment of sites and organisations
- 14,031: Environmental performance evaluation
- 14,050: Vocabulary
- 14,040, 44, 47, and 49: Life cycle assessment
- 14,062: Integrating environmental aspects into product design and development
- 14,063: Environmental communication.

An ISO certification can be costly, between US$5,000 and 20,000 for the first audit and an annual cost of between US$4,000 and 5,000 (NGLS & UNRISD, 2002), while implementation can incur a long adaptation process (DeMendonça & Baxter, 2001).

EMAS is a management tool for companies and other organisations to evaluate, report, and improve their environmental performance. It follows the same principles as ISO14000 (i.e. conducting environmental reviews, establishing an EMS, carrying audits, and providing a statement of company environmental performance). Its registration fees can vary from zero to 1500 dollars in the case of large companies. The main differences with ISO 14000 is that EMAS is European (whereas ISO 14000's scope is worldwide) and that EMAS sets stricter requirements in some areas (Brorson & Larsson, 1999). The EMAS III has improved registration procedures, applicability for small and medium sized enterprises, and registration of several sites (which can be worldwide) under one corporation (EC, 2011).

4.5.3 Integrated Management Systems

Integrated management systems (IMS) is an approach to managing processes or activities that transform resource inputs into a product or service which meet an organisation's objectives and equitably satisfy the requirements of stakeholders, quality, health, safety, environmental, security, ethical, or any other identified requirement (Jørgensen et al., 2006; Olaru et al., 2014). IMS is designed to engage with stakeholders and cater to their needs (Asif et al., 2013). IMS is considered the best management practice when an organisation has multiple MSs in place (Bernardo, 2014).

4.5.4 Sustainability Reporting (SR)

Sustainability reporting (SR) is a voluntary activity with the following purposes: to (1) assess the current state of an organisation's progress towards sustainability; (2) communicate the efforts and progress in the economic, environmental, and social dimensions to stakeholders (Dalal-Clayton & Bass, 2002; GRI, 2011); (3) assess sustainability performance over time; (4) benchmark against other companies; (5) demonstrate how the organisation influences, and is influenced by, expectations about sustainable development (Daub 2007; Schaltegger & Wagner, 2006); and (6) be a base for planning changes for sustainability (Adams & McNicholas, 2007; Lozano, 2013a; Lozano, Llobet, and Tideswell 2013).

Various European countries have introduced regulations concerning corporate SR, with a view to lowering the costs entailed by stakeholders in reducing the information asymmetry between themselves and the firms (Herzig & Schaltegger, 2006), particularly through EU Directive 2014/95/EU on non-financial reporting (European Parliament, 2014). In general, European companies have been at the forefront of sustainability reporting (Kolk, 2008; Lozano, 2013a).

During the last decade, SR has been increasingly recognised as a key driver of sustainability in organisations (Ceulemans et al., 2015; Domingues et al., 2017; Lozano et al., 2016; Morhardt et al., 2002). SR can be driven by (1) only inside motivations, (2) mainly inside motivations, (3) equal inside motivations and external pressures, (4) mainly external pressures, and (5) only external pressures (Ceulemans et al., 2015; Domingues et al., 2017; Lozano et al., 2016).

There are three approaches to assessing and reporting sustainability that may be used independently or in combination (Dalal-Clayton & Bass, 2002):

1. Accounts: construction of raw data that are then converted to a common unit, e.g. monetary, area, or units of energy.
2. Narrative assessments: combinations of text, maps, graphics, and tabular data. Narrative assessments might use indicators, but they are not a cornerstone.
3. Indicator-based: these may include texts, maps, graphics, and tabular data, as with the narrative assessment, but they are organised around indicators.

Table 4.3 Three main approaches to measure and analyse Sustainability. Source: (Dalal-Clayton & Bass, 2002)

Approach potential	Accounts	Narrative assessments	Indicator-based assessments
1. Transparency	Low	Medium	High
2. Consistency	High	Low	High
3. Participation	Low	High	Medium
4. Usefulness for decision-making	Medium	Medium	High

Each type of report offers advantages and disadvantages over the others, as presented in Table 4.3.

SR presents a number of challenges, such as gaining knowledge, experience, and understanding of sustainability (Adams & McNicholas, 2007). Lozano (2006, 2013a) recognised the following inherent disadvantages of sustainability assessment and reporting: once started, the process of reporting becomes almost impossible to stop, stakeholders tend to demand more from the corporation/institution, keeping up the balance between details and core information presents a big challenge, extra resources and time are needed to gather all the data to fulfil the indicators and to engage the stakeholders, and "reporting fatigue".

Three of the latest development in SR have been the (1) development of a framework combining financial, social, and environmental information (ICCA 2011; IIRC, 2011), (2) development of integrated or interlinking indicators (Azapagic, 2004; Lozano, 2013a; Lozano & Huisingh, 2011), and (3) assessment of performance (e.g. Daub 2007; Lozano, 2006, 2013a; Skouloudis & Evangelinos, 2009).

A large number of standards and guidelines have been developed during the last two decades (Lozano & Huisingh, 2011; Perrini & Tencati, 2006). The most widely used include the **ISO14000** series and EMAS, covered in the EMS section, **AA1000** Stakeholder Engagement Standard from Accountability, Social Accountability 8000 Standard (SAI, 2007), **ISO 26000** (Hahn & Weidtmann, 2016; Schwartz & Tilling, 2009), and **GRI Sustainability Guidelin**es (Kuehr, 2007).

The **AA1000** helps to establish a systematic stakeholder engagement process to ensure greater transparency and effective responsiveness to stakeholders (ISEA, 1999). It involves stakeholder management throughout the entire process (Lozano & Huisingh, 2011). Its emphasis is on innovation over compliance and the possibility to chart their own course as opposed to being guided (Leipziger, 2003).

SA8000 is an auditable certification standard based on international workplace norms of International Labour Organisation (ILO) conventions, the Universal Declaration of Human Rights, and the UN Convention on the Rights of the Child (SAI, 2007). It addresses human and labour rights explicitly throughout the company (Lozano & Huisingh, 2011). It raises public awareness about the company's efforts (SAI, 2007).

ISO 26000 was developed for the purpose of giving guidance on the social responsibility of organisations (Hahn & Weidtmann, 2016; Schwartz & Tilling, 2009). ISO 26000 is an important step in improving sustainability performance

(Hahn, 2013), since it focuses on the needs of its direct users and also on the needs of all groups in society (Hahn & Weidtmann, 2016). ISO 26000 has to date only been in use for a limited time (Hahn, 2013; Hahn & Weidtmann, 2016).

The **GRI Sustainability Guidelines** are one of the most complete and worldwide recognised guidelines available (Hussey et al., 2001; Lozano, 2006; Roca & Searcy, 2012). They are voluntary guidelines for reporting on sustainability issues (Global Reporting Initiative (GRI), 2018). The GRI guidelines consist of a set of, usually compartmentalised (Lozano & Huisingh, 2011), topic-specific standards with indicators covering the economic, environmental, and social dimensions (GRI, 2016). A GRI report is developed through two main stages (collecting data and populating the indicators) (Lozano, Llobet, and Tideswell 2013). These can be complemented by a third stage, assessing coverage and performance, to provide a quantitative and comparative perspective on where a report is performing well and where it could be improved (Lozano, 2006, 2013a).

4.6 TIA 'Discussion'

The majority of the TIAs have focused on the economic and environmental dimensions (as discussed by Atkinson, 2000; Lozano, 2012; Reinhardt, 2000) and on technocentric and managerial ploys (Lozano, 2015). The TIAs have been limited in addressing all the system elements of organisations, since they have not been decided, nor used, to address explicitly the governance, organisational systems, service provision, or change management system elements.

Each TIA has advantages with respect to the sustainability dimensions and its contribution to an organisation's system elements, but it also has limitations in its scope and reach (Lozano, 2020). Relying on one initiative can result in a limited and narrow contribution to sustainability and curtail coverage of the company's system, while using too many tools wastes resources and energy due to duplication in tasks (Lozano, 2012).

None of the TIAs, on its own, covers the full organisation system or the four dimensions of sustainability (economic, environmental, social, and time) (Lozano 2012). A combination of TIAs is needed, which capitalises on synergies and address the sustainability dimensions and the organisation's system elements, where between four and six TIAs are considered to be the optimal combination (Lozano, 2012, 2020). An example of such combination is the Corporate and Industrial Voluntary Initiatives for Sustainability (CIVIS) framework (see Table 4.4) aimed at (1) helping company leaders better understand how to improve their company's contribution to sustainability and (2) fostering a holistic approach through the combination of TIAs to embed sustainability into a company's system with the least effort and maximum results (Lozano, 2012).

Table 4.4 Corporate and Industrial Voluntary Initiatives for Sustainability (CIVIS) framework. Source: colour added to (Lozano, 2012)

Corporate initiative	Corporate System					Sustainability			
	Operations and Production	Management and Strategy	Organisational Systems	Procurement and Marketing	Assessment and Reporting	Economic.	Environmental	Social	Time
Sustainable Livelihoods		Full				Full	Full	Full	Full
Triple Bottom Line		Full		Full		Full	Full	Full	
The Natural Step		Full	Limited			Limited	Full	Limited	Limited
Environmental Management Systems		Full	Limited		Full		Full		
Environmental and Social Accounting		Full			Full	Full	Full	Full	
Life Cycle Assessment	Full				Full		Full		
Cleaner Production	Full					Full	Full		
Design for Environment	Full						Full		
Eco-efficiency	Full					Full	Full		
Industrial Ecology	Full					Full	Full		
Factor X	Full					Limited	Full		
Green Chemistry	Full						Full		
Eco-labelling				Full	Full		Full		
Corporate Social Responsibility		Full			Limited		Limited	Full	
Sustainability Reporting		Limited			Full	Full	Variable	Variable	
Corporate Citizenship		Full						Full	

Colour coding
- Full contribution (dark green)
- Limited contribution (light green)
- Variable contribution (orange)

References

Abdelkafi, N., & Tauscher, K. (2015). Business models for sustainability from a system dynamics perspective. *Organization & Environment, 1–23*. https://doi.org/10.1177/1086026615592930

Adams, C. A., & McNicholas, P. (2007). Making a difference. Sustainability reporting, accountability and organisational change. *Accounting, Auditing & Accountability Journal, 20*(3), 382–402. https://doi.org/10.1108/09513570710748553

Ahi, P., & Searcy, C. (2013). A comparative literature analysis of definitions for green and sustainable supply chain management. *Journal of Cleaner Production, 52*, 329–341. https://doi.org/10.1016/j.jclepro.2013.02.018

Anastas, P. T. (2000). Green chemistry. In *Theory and practice*. Oxford University Press.

Anastas, P. T., & Warner, J. C. (1998). Green chemistry. In *Theory and practice*. Oxford University Press.

Anastas, Paul T, & Breen, J. J. (1997). Design for the environment and green chemistry: The heart and soul of industrial ecology. Journal of Cleaner Production, 5(1–2), 97–102.

Andersen, M. S. (2006). An introductory note on the environmental economics of the circular economy. *Sustainability Science, 2*(1), 133–140. https://doi.org/10.1007/s11625-006-0013-6

Aras, G., & Crowther, D. (2008). Governance and sustainability: An investigation into the relationship between corporate governance and corporate sustainability. *Management Decision, 46*(3), 433–448. https://doi.org/10.1108/00251740810863870

Ashley, S. (1993). Designing for the environment. *Mechanical Engineering, 115*(3), 55.

Asif, M., Searcy, C., Zutshi, A., & Fisscher, O. A. M. (2013). An integrated management systems approach to corporate social responsibility. *Journal of Cleaner Production, 56*, 7–17. https://doi.org/10.1016/j.jclepro.2011.10.034

Atkinson, G. (2000). Measuring corporate sustainability. *Journal of Environmental Planning and Management, 43*(2), 235–252.

Avi-Yonah, R. (2005). The cyclical transformations of the Coporate form: A historical Pespective on corporate social responsibility. *Delaware Journal of Corporate Law, 30*, 767–818. http://law.bepress.com/umichlwps-olin/art38/

Azapagic, A. (2004). Developing a framework for sustainable development indicators for the mining and minerals industry. *Journal of Cleaner Production, 12*, 639–662.

Barreiro-Gen, M., & Lozano, R. (2020). How circular is the circular economy? Analysing the implementation of circular economy in organisations. *Business Strategy and the Environment, 29*(8), 3484–3494. https://doi.org/10.1002/bse.2590

Bartelmus, P. (1999). *Sustainable Development - Paradigm or paranoia?* Wuppertal Institute.

Focusing capital on the long term. (2014). *January-Fe Harvard Business Review*. https://doi.org/10.1353/abr.2012.0147

Baumgartner, R. J., & Zielowski, C. (2007). Analyzing zero emission strategies regarding impact on organizational culture and contribution to sustainable development. *Journal of Cleaner Production, 15*, 1321–1327.

Berle, A. A., Berle, A. A., Jr., Berle, A. A., & Berle, A. A., Jr. (1947). The theory of enterprise entity. *Columbia Law Review, 47*(3), 343–358.

Bernardo, M. (2014). Integration of management systems as an innovation: A proposal for a new model. *Journal of Cleaner Production, 82*, 132–142. https://doi.org/10.1016/j.jclepro.2014.06.089

Bilitewski, B. (2012). The circular economy and its risks. *Waste Management (New York, N.Y.), 32*(1), 1–2. https://doi.org/10.1016/j.wasman.2011.10.004

Birch, D., & Littlewood, G. (2004). Corporate citizenship. Some perspectives from Australian CEOs. *Journal of Corporate Citizenship, 16*, 61–69.

Bocken, N. M. P. M. P., Short, S. W. W., Rana, P., & Evans, S. (2014). A literature and practice review to develop sustainable business model archetypes. *Journal of Cleaner Production, 65*, 42–56. https://doi.org/10.1016/j.jclepro.2013.11.039

Bohnsack, R., Pinkse, J., & Kolk, A. (2014). Business models for sustainable technologies: Exploring business model evolution in the case of electric vehicles. *Research Policy, 43*(2), 284–300. https://doi.org/10.1016/j.respol.2013.10.014

Boons, F., & Lüdeke-Freund, F. (2013). Business models for sustainable innovation: State-of-the-art and steps towards a research agenda. *Journal of Cleaner Production, 45*, 9–19. https://doi.org/10.1016/j.jclepro.2012.07.007

Borchardt, M., Wendt, M. H., Pereira, G. M., & Sellitto, M. A. (2011). Redesign of a component based on ecodesign practices: Environmental impact and cost reduction achievements. *Journal of Cleaner Production, 19*(1), 49–57. https://doi.org/10.1016/j.jclepro.2010.08.006

Bovea, M. D., & Pérez-Belis, V. (2012). A taxonomy of ecodesign tools for integrating environmental requirements into the product design process. *Journal of Cleaner Production, 20*(1), 61–71. https://doi.org/10.1016/j.jclepro.2011.07.012

Brorson, T., & Larsson, G. (1999). *Environmetal management* (Third). Abrahamsons Tryckeri AB.

Burritt, R. L., Herzig, C., & Tadeo, B. D. (2009). Environmental management accounting for cleaner production: The case of a Philippine rice mill. *Journal of Cleaner Production, 17*(4), 431–439. https://doi.org/10.1016/j.jclepro.2008.07.005

C.E.C. (2001). *Promoting a European framework for corporate social responsibility*. Commission of the European Communities.

C.E.C. (2002). *Corporate social responsibility: A business contribution to sustainable development*. Commission of the European Communities.

Carpenter, A., Lozano, R., Sammalisto, K., & Astner, L. (2018). Securing a port's future through circular economy: Experiences from the port of Gävle in contributing to sustainability. *Marine Pollution Bulletin, 128*, 539–547. https://doi.org/10.1016/j.marpolbul.2018.01.065

Carroll, A. B. (1998). The four faces of corporate citizenship. *Business and Society Review, 100*(101), 1–7.

Carroll, A. B. (1999). Corporate social responsibility: Evolution of a definitional construct. *Business Society, 38*(3), 268–295.

Ceulemans, K., Lozano, R., & Alonso-Almeida, M. (2015). Sustainability reporting in higher education: Interconnecting the reporting process and Organisational change Management for Sustainability. *Sustainability, 7*(7), 8881–8903. https://doi.org/10.3390/su7078881

Chambers, R. (1995). Poverty and livelihoods: Whose reality counts? *Environment and Urbanization, 7*(1), 173–203.

Chava, S. (2010). Socially responsible investing and expected stock returns. *SSRN Electronic Journal*. https://doi.org/10.2139/ssrn.1678246

Cohen, B., & Kietzmann, J. (2014). Ride on! Mobility business models for the sharing economy. *Organization & Environment, 27*(3), 279–296. https://doi.org/10.1177/1086026614546199

Dalal-Clayton, B., & Bass, S. (2002). *Sustainable Development Strategies: A Resource Book* (Organization for Economic Co-operation and Development (ed.); Paris and). Earthscan Publications Ltd.

Daub, C.-H. (2007). Assessing the quality of sustainability reporting: An alternative methodological approach. *Journal of Cleaner Production, 15*(1), 75–85. https://doi.org/10.1016/j.jclepro.2005.08.013

de Beer, P., & Friend, F. (2006). Environmental accounting: A management tool for enhancing corporate environmental and economic performance. *Ecological Economics, 58*(3), 548–560. https://doi.org/10.1016/j.ecolecon.2005.07.026

DeMendonça, M., & Baxter, T. E. (2001). Design for the environment (DFE): An approach to achieve the ISO 14000 international standardization. *Environmental Management and Health, 12*(1), 51–56.

DeSimone, L. D., & Popoff, F. (2000). *Eco-Efficiency*. MIT Press.

Dodd, E. M., Jr. (1932). For whom are corporate managers trustees? *Harvard Law Review, XLV*(7), 1145–1163.

Domingues, A. R., Lozano, R., Ceulemans, K., & Ramos, T. B. (2017). Sustainability reporting in public sector organisations: Exploring the relation between the reporting process and organisational change management for sustainability. *Journal of Environmental Management, 192*, 292–301. https://doi.org/10.1016/j.jenvman.2017.01.074

Doppelt, B. (2003). Overcoming the seven sustainability blunders. *The Systems Thinker, 14*(5), 2–7.

Dunphy, D., Griffiths, A., & Benn, S. (2003). *Organizational change for corporate sustainability* (Third). Routledge.

Dyllick, T., & Hockerts, K. (2002). Beyond the business case for corporate sustainability. *Business Strategy and the Environment, 11*, 130–141.

EC. (2009). *Directive 2009/125/EC of the European Parliaments and the Council of 21 October 2009 establishing a framework for the setting of ecodesign requirements for energy-related products (recast)* (Vol. 2011, Issue 7 June). European Commission Energy. http://eur-lex.europa.eu/LexUriServ/LexUriServ.do?uri=OJ:L:2009:285:0010:0035:EN:PDF

EC. (2011). *EMAS factsheet*. European Commission. http://ec.europa.eu/environment/emas/pdf/factsheet/fs_EMAS2to3.pdf

Ehrenfeld, J. (2004). Industrial ecology: A new field or only a metaphor? *Journal of Cleaner Production, 12*(8–10), 825–831. https://doi.org/10.1016/j.jclepro.2004.02.003

Ehrenfeld, J., & Gertler, N. (1997). Industrial ecology in practice. The evolution of interdependence at Kalundborg. *Journal of Industrial Ecology, 1*(1), 67–79.

Ehrenfeld, J. R. (2005). Eco-efficiency. Philosophy, theory, and tools. *Journal of Industrial Ecology, 9*(4), 6–8.

Ekins, P. (2005). Eco-efficiency. Motives, drives, and economic implications. *Journal of Industrial Ecology, 9*(4), 12–14.

Elkington, J. (1998). *Cannibals with forks: The triple bottom line of 21st century business.* New Society Publishers.

Elkington, J. (2002). Cannibals with forks. The triple bottom line of the 21st century business. In *Conscientious Commerce*. Capstone Publishing Co.. http://doi.wiley.com/10.1002/tqem.3310080106

Escrig-olmedo, E., Rivera-lirio, J. M., Jesús, M., & Angeles, M. (2017). Integrating multiple ESG investors ' preferences into sustainable investment: A fuzzy multicriteria methodological approach 162, 1334–1345. https://doi.org/10.1016/j.jclepro.2017.06.143.

EU. (2003). Waste electrical and electronic equipment. EU.

European Commision. (2020). Circular Economy. https://ec.europa.eu/growth/industry/sustainability/circular-economy_en.

European Commission. (2014a). The circular economy: Connecting, creating and conserving value. https://doi.org/10.2779/80121.

European Commission. (2014b). Towards a circular economy: A zero waste programme for Europe.

European Commission. (2018). A monitoring framework for the circular economy. COM(2018) 29 final.16.1.2018. In *COM/2018/29 final* (Vol. 29, Issue final). COM/2018/029 final.

European Commission, & European Commission – Environment. (2012). REACH. European Commission. http://ec.europa.eu/environment/chemicals/reach/reach_intro.htm

European Parliament. (2014). Directive 2014/95/EU of the European Parliament and of the council of 22 October 2014 amending directive 2013/34/EU as regards disclosure of non-financial and diversity information by certain large undertakings and groups.

Eurosif. (2014). *European SRI Study*, 2014.

Farmer, R. N., & Hogue, W. D. (1973). *Corporate social responsibility*. Science Research Associate.

Fernández, I., & Kekäle, T. (2005). The influence of modularity and industry clockspeed on reverse logistics strategy: Implications for the purchasing function. *Journal of Purchasing and Supply Management, 11*(4), 193–205. https://doi.org/10.1016/j.pursup.2006.01.005

Finkbeiner, M., Inaba, A., Tan, R., Christiansen, K., & Klüppel, H.-J. (2006). The new international standards for life cycle assessment: ISO 14040 and ISO 14044. *The International Journal of Life Cycle Assessment, 11*(2), 80–85. https://doi.org/10.1065/lca2006.02.002

Frankental, P. (2001). Corporate social responsibility - a PR invention? *Corporate Communications: An International Journal, 6*(1), 18–23.

Frehs, J. (2003). *Corporate social responsibility: Lessons learned.* Natural Resources Canada.

Fukukawa, K., & Moon, J. (2004). A Japanese model of corporate social responsibility? *Journal of Corporate Citizenship, 16*, 45–59.

Galarraga Gallastegui, I. (2002). The use of eco-labels: A review of the literature. *European Environment, 12*, 316–331.

Gerrard, J., & Kandlikar, M. (2007). Is European end-of-life vehicle legislation living up to expectations? Assessing the impact of the ELV directive on "green" innovation and vehicle recovery. *Journal of Cleaner Production, 15*, 17–27. https://doi.org/10.1016/j.jclepro.2005.06.004

Gibbs, D., & Deutz, P. (2007). Reflections on implementing industrial ecology through eco-industrial park development. *Journal of Cleaner Production, 15*, 1683–1695.

Glavič, P., & Lukman, R. (2007). Review of sustainability terms and their definitions. *Journal of Cleaner Production, 15*(18), 1875–1885. https://doi.org/10.1016/j.jclepro.2006.12.006

Global Reporting Initiative (GRI). (2018). *Sustainability disclosure database*. GRI.

Gompers, P. A., Ishii, J. L., & Metrick, A. (2003). Corporate governance and equity prices. *Quarterly Journal of Economics, 118*(February), 107–155. https://doi.org/10.1162/00335530360535162

GRI. (2011). *Sustainability Reporting Guidelines 3.1*. Global Reporting Initiative. http://www.globalreporting.org/NR/rdonlyres/53984807-9E9B-4B9F-B5E8-77667F35CC83/0/G31GuidelinesinclTechnicalProtocolFinal.pdf

de Haes, H. A. U., Jolliet, O., Norris, G., & Saur, K. (2002). UNEP/SETAC life cycle initiative: Background, aims and scope. *The International Journal of Life Cycle Assessment, 7*(4), 192–195. https://doi.org/10.1007/BF02978872

Hahn, R. (2013). ISO 26000 and the standardization of strategic management processes for sustainability and corporate social responsibility. *Business Strategy and the Environment, 22*(7), 442–455. https://doi.org/10.1002/bse.1751

Hahn, R., & Weidtmann, C. (2016). Transnational governance, deliberative democracy, and the legitimacy of ISO 26000. *Business & Society, 55*(1), 90–129. https://doi.org/10.1177/0007650312462666

Hale, M. (1996). Ecolabelling and cleaner production: Principles, problems, education and training in relation to the adoption of environmentally sound production processes. *Journal of Cleaner Production, 4*(2), 85–95.

Hallstedt, S. (2008). A foundation for sustainable product development. In *Mechanical Engineering: Vol.* PhD. Blekinge Institute of Technology.

Hamann, R. (2003). Mining companies' role in sustainable development: The "why" and "how" of corporate social responsibility from a business perspective. *Development Southern Africa, 20*(2), 234–254.

Heeres, R., Vermeulen, W. J. V., & de Walle, F. B. (2004). Eco-industrial park initiatives in the USA and the Netherlands: First lessons. *Journal of Cleaner Production, 12*, 985–995.

Heine, L. (2007). Sustainable materials and green chemistry. In McGraw-Hill (Ed.), Access science. Encyclopedia of Science & Technology Online. McGraw-Hill. DOI https://doi.org/10.1036/1097-8542.YB030505.

Heiskanen, E. (2002). The institutional logic of life cycle thinking. *Journal of Cleaner Production, 10*(5), 427–437. https://doi.org/10.1016/S0959-6526(02)00014-8

Henderson, D. (2005). The role of business in the world of today. *Journal of Corporate Citizenship, 17*, 30–32.

Herzig, C., & Schaltegger, S. (2006). Corporate sustainability reporting: An overview. In S. Schaltegger, M. Bennet, & R. L. Burrit (Eds.), *Sustainability accounting and reporting*. Springer.

Hockerts, K. (1999). The sustainability radar - a tool for the innovation of sustainable products and services. *Greener Management International, 25*, 29–49.

Holliday, C. O. J., Schmidheiny, S., & Watts, P. (2002). Walking the talk. In *The Business Case for Sustainable Development*. Greenleaf Publishing.

Holmberg, J., & Robert, K.-H. (2000). Backcasting: A framework for strategic planning. *International Journal of Sustainable Development and World Ecology, 7*(4), 291–308.

Holme, R., & Watts, P. (2000). *Corporate social responsibility: Making good business sense*. WBCSD.

Hopkins, M. J. D. (2002). Sustainability in the internal operations of companies. *Corporate Environmental Strategy, 9*(2), 1–11.

Hussey, D. M., Kirsop, P. L., & Meissen, R. E. (2001). Global reporting initiative guidelines: An evaluation of sustainable Development metrics for industry. *Environmental Quality Management*, 1–20.

ICCA. (2011). *Responsible care*. International Council of Chemical Associations. http://www.icca-chem.org/en/Home/Responsible-care/

IIRC. (2011). Towards Integrated Reporting: Communicating Value in the 21st Century. International Integrated Reporting Committee.

ISEA. (1999). AA1000 - Standards, Guidelines and Professional Qualification.

Isenmann, R. (2003). Industrial ecology: Shedding more light on its perspective of understanding nature as model. *Sustainable Development, 11*, 143–158.
ISO. (2021). *ISO 14000 Family Environmental Management*. ISO.
Ite, U. E. (2004). Multinationals and corporate social responsibility in developing countries: A case suty of Nigeria. *Corporatare Social Responsiblity and Environmental Management, 11*, 1–11.
Jacobsen, N. B. (2006). Industrial Symbiosis in Kalundborg, Denmark. A quantitative assesment of economic and environmental aspects. *Journal of Industrial Ecology, 10*(1–2), 239–255.
Jakl, T., & Schwager, P. (2008). Lessons learned – The way forward. In *Chemical leasing goes global: Selling services instead of barrels – A win-win business model for environment and industry* (pp. 221–226). Springer.
Jansen, L. (2003). The challenge of sustainable development. *Journal of Cleaner Production, 11*, 231–245.
Jenkins, R. (2002). Corporate codes of conduct: Self-regulation in a global economy. In NGLS & UNRISD (Ed.), *Voluntary approaches to corporate responsibility. Readings and a resource guide*. UN Non-Governmental Liaison Service.
Jensen, M. M., & Meckling, W. H. (1976). Theory of the firm: Managerial behavior, agency costs and ownership structure. *Journal of Financial Economics, 3*(4), 305–360.
Johnson, M., Redlbacher, F., & Schaltegger, S. (2018). Stakeholder engagement for corporate sustainability: A comparative analysis of B2C and B2B companies. *Corporate Social Responsibility and Environmental Management*. https://doi.org/10.1002/csr.1484
Jørgensen, T. H., Remmen, A., & Mellado, M. D. (2006). Integrated management systems – Three different levels of integration. *Journal of Cleaner Production, 14*(8), 713–722. https://doi.org/10.1016/j.jclepro.2005.04.005
King, A. M., Burgess, S. C., Ijomah, W., & McMahon, C. A. (2006). Design for end-of-Life: Repair, recondition, remanufacture or recycle? *Sustainable Development, 14*, 257–267. https://doi.org/10.1115/DETC2004-57759
Klettner, A., Clarke, T., & Boersma, M. (2014). The Governance of Corporate Sustainability: Empirical Insights into the Development, Leadership and Implementation of Responsible Business Strategy. *Journal of Business Ethics, 122*(1), 145–165. https://doi.org/10.1007/s10551-013-1750-y
Kolk, A. (2008). Sustainability, accountability and corporate governance: Exploring multinationals' reporting practices. *Business Strategy and the Environment, 18*, 1–15.
Korhonen, J. (2003). Should we measure corporate social responsibility? *Corporate Social Responsibility and Environmental Management, 10*, 25–39.
Kotler, P., & Armstrong, G. (2001). *Principles of marketing* (ninth). Prentice Hall.
Kromer, M. A., Joseck, F., Rhodes, T., Guernsey, M., & Marcinkoski, J. (2009). Evaluation of a platinum leasing program for fuel cell vehicles. *International Journal of Hydrogen Energy, 34*, 8276–8288. https://doi.org/10.1016/j.ijhydene.2009.06.052
Kuehr, R. (2007). Towards a sustainable society: United Nations University's zero emissions approach. *Journal of Cleaner Production, 15*, 1198–1204.
Laffer, A. B., Coors, A., & Winegarden, W. (2004). *Does corporate social responsibility enhance business profitability?* Laffer Associates.
Langer, M. E., & Schön, A. (2003). Enhancing corporate sustainability. In *A framework based evaluation tools for sustainable development*. Forschungsschwerpunkt Nachhaltigkeit und Umweltmanagement, Wirtschaftsuniversität Wien.
Lantos, G. (2001). The boundaries of strategic corporate social responsibility. *Journal of Consumer Marketing, 18*(7), 595–630.
Lay, G., Schroeter, M., & Biege, S. (2009). Service-based business concepts: A typology for business-to-business markets. *European Management Journal, 27*, 442–455. https://doi.org/10.1016/j.emj.2009.04.002
Le Pochat, S., Bertoluci, G., & Froelich, D. (2007). Integrating ecodesign by conducting changes in SMEs. *Journal of Cleaner Production, 15*(7), 671–680. https://doi.org/10.1016/j.jclepro.2006.01.004

Lehman, G. (1999). Disclosing new worlds: A role for social and environmental accounting and auditing. *Accounting, Organizations and Society, 24*(3), 217–241. https://doi.org/10.1016/S0361-3682(98)00044-0

Leipziger, D. (2003). *The corporate responsibility: Code book*. GreenLeaf Publishing.

Leisinger, K. M. (2003). Opportunities and risks of the United Nations global compact. *Journal of Corporate Citizenship, 11*(113–130).

Leontief, W. (1928). Die Wirtschaft als Kreislauf..

Leontief, W. (1991). The economy as a circular flow. *Structural Change and Economic Dynamics, 2*(1).

Linnenluecke, M. K., Russell, S. V., Griffi, A., Linnenluecke, M. K., & Griffiths, A. (2009). Subcultures and sustainability practices : The impact on understanding corporate sustainability. *Business Strategy and the Environment, 18*(December 2007), 432–452. https://doi.org/10.1002/bse

Liu, L., Liang, Y., Song, Q., & Li, J. (2017). A review of waste prevention through 3R under the concept of circular economy in China. *Journal of Material Cycles and Waste Management, 19*(4), 1314–1323. https://doi.org/10.1007/s10163-017-0606-4

Lowe, E. A., & Evans, L. K. (1995). Industrial ecology and industrial ecosystems. *Journal of Cleaner Production, 3*(1–2), 47–53.

Lowenthal, M. D., & Kastenberg, W. E. (1998). Industrial ecology and energy systems: A first step. *Resources, Conservation and Recycling, 24*(51–63).

Lozano, R. (2006). A tool for a graphical assessment of sustainability in universities (GASU). *Journal of Cleaner Production, 14*(9–11), 963–972. https://doi.org/10.1016/j.jclepro.2005.11.041

Lozano, R. (2008). Envisioning sustainability three-dimensionally. *Journal of Cleaner Production, 16*(17), 1838–1846. https://doi.org/10.1016/j.jclepro.2008.02.008

Lozano, R. (2012). Towards better embedding sustainability into companies' systems: An analysis of voluntary corporate initiatives. *Journal of Cleaner Production, 25*(0), 14–26. https://doi.org/10.1016/j.jclepro.2011.11.060

Lozano, R. (2013a). Sustainability inter-linkages in reporting vindicated: A study of European companies. *Journal of Cleaner Production, 51*, 57–65. https://doi.org/10.1016/j.jclepro.2013.01.039

Lozano, R. (2013b). Are companies planning their Organisational changes for corporate sustainability? An analysis of three case studies on resistance to change and their strategies to overcome it. *Corporate Social Responsibility and Environmental Management, 20*(5), 275–295. https://doi.org/10.1002/csr.1290

Lozano, R. (2015). A holistic perspective on corporate sustainability drivers. *Corporate Social Responsibility and Environmental Management, 22*(1), 32–44. https://doi.org/10.1002/csr.1325

Lozano, R. (2018). Sustainable business models: Providing a more holistic perspective. *Business Strategy and the Environment, 27*(8), 1159–1166. https://doi.org/10.1002/bse.2059

Lozano, R. (2020). Analysing the use of tools, initiatives, and approaches to promote sustainability in corporations. *Corporate Social Responsibility and Environmental Management, 27*(2), 982–998. https://doi.org/10.1002/csr.1860

Lozano, R., Bautista-Puig, N., & Barreiro-Gen, M. (2021). Elucidating a holistic and panoptic framework for analysing circular economy. *Business Strategy and the Environment, 30*(4), 1644–1654. https://doi.org/10.1002/bse.2699

Lozano, R., Carpenter, A., & Lozano, F. J. (2014). Critical reflections on the chemical leasing concept. *Resources, Conservation and Recycling, 86*, 53–60. https://doi.org/10.1016/j.resconrec.2014.02.003

Lozano, R., Carpenter, A., & Satric, V. (2013). Fostering green chemistry through a collaborative business model: A chemical leasing case study from Serbia. *Resources, Conservation and Recycling, 78*(November), 136–144. https://doi.org/10.1016/j.resconrec.2013.07.007

Lozano, R., & Huisingh, D. (2011). Inter-linking issues and aspects in sustainability reporting. *Journal of Cleaner Production, 19*(2–3), 99–107. https://doi.org/10.1016/j.jclepro.2010.01.004

Lozano, R., Llobet, J., & Tideswell, G. (2013). The process of assessing and reporting sustainability at universities: Preparing the report of the University of Leeds. *Sostenibilidad, Tecnologia y Humanismo, 6*, 85–112.

Lozano, R., Nummert, B., & Ceulemans, K. (2016). Elucidating the relationship between sustainability reporting and Organisational change Management for Sustainability. *Journal of Cleaner Production, 125*, 168–188. https://doi.org/10.1016/j.jclepro.2016.03.021

Lozano, R., & Reid, A. (2017). Socially responsible or reprehensible? Investors, electricity utility companies, and transformative change in Europe. *Energy Research & Social Science, 37*(September 2017), 37–43. https://doi.org/10.1016/j.erss.2017.09.018

Lozano, R., Suzuki, M., Carpenter, A., & Tyunina, O. (2017). An analysis of the contribution of Japanese business terms to corporate sustainability: Learnings from the "looking-glass" of the east. *Sustainability (Switzerland), 9*(2), 1–17. https://doi.org/10.3390/su9020188

MacLeod, S., & Lewis, D. (2004). Transnational corporations. Power, influence and responsibility. *Global Social Policy, 4*(I), 77–98.

Matten, D., & Moon, J. (2004). Corporate social responsibility education in Europe. *Journal of Business Ethics, 54*, 323–337.

Matus, K. J. M. (2010). Policy incentives for a cleaner supply chain: The case of green chemistry. *Journal of International Affairs, 64*(1), 121–136.

Matus, K. J. M., Clark, W. C., Anastas, P. T., & Zimmerman, J. B. (2012). Barriers to the implementation of green chemistry in the United States. *Environmental Science & Technology, 46*, 10892–10899. https://doi.org/10.1021/es3021777

McDaniel, S. W., & Rylander, D. H. (1993). Strategic green marketing. *Journal of Consumer Marketing, 10*(3), 4–10. https://doi.org/10.1108/07363769310041929

McIntosh, M., Leipziger, D., Jones, K., & Coleman, G. (1998). Corporate citizenship. In *Successful strategies for responsible companies*. Financial Times. Pitman Publishing.

Millon, D. (1990). Theories of the corporation. *Duke Law Journal, 39*(2), 201–262. https://scholarship.law.duke.edu/dlj/vol39/iss2/2

Milne, M. J., Kearins, K., & Walton, S. (2003). Business makes a "journey" out of "sustainability": Creating adventures in wonderland?

Mintzberg, H. (1983). The case for corporate social responsibility. *Journal of Business Strategy, 4*(2), 3–13.

Mont, O., Dalhammar, C., & Jacobsson, N. (2006). A new business model for baby prams based on leasing and product remanufacturing. *Journal of Cleaner Production, 14*, 1509–1518. https://doi.org/10.1016/j.jclepro.2006.01.024

Morhardt, J. E. E., Baird, S., & Freeman, K. (2002). Scoring corporate environmental and Sustainaibility reports using GRI 2000, ISO 14031 and other criteria. *Corporate Social Responsibility and Environmental Management, 9*, 215–233.

Murray, A., Skene, K., & Haynes, K. (2015). The circular economy: An interdisciplinary exploration of the concept and application in a global context. *Journal of Business Ethics*. https://doi.org/10.1007/s10551-015-2693-2

Nadaı, A. (1999). Conditions for the development of a product ecolabel. *European Environment, 9*, 202–211.

NGLS, & UNRISD. (2002). Voluntary approaches to corporate responsibility. In *Readings and a resource guide*. UN Non-Governmental Liaison Service.

OECD. (1997). *Eco-labelling: Actual effects of selected programmes*. Organisation for Economic Co-operation and Development.

Olaru, M., Maier, D., Nicoară, D., & Maier, A. (2014). Establishing the basis for Development of an organization by adopting the integrated management systems: Comparative study of various models and concepts of integration. *Procedia - Social and Behavioral Sciences, 109*, 693–697. https://doi.org/10.1016/j.sbspro.2013.12.531

Parker, L. D. (2011). Twenty-one years of social and environmental accountability research: A coming of age. *Accounting Forum, 35*(1), 1–10. https://doi.org/10.1016/j.accfor.2010.11.001

Pauli, G. (1997). Zero emissions: The ultimate goal of cleaner production. *Journal of Cleaner Production, 5*(1–2), 109–113. https://doi.org/10.1016/s0959-6526(97)00013-9

Peattie, K. (1995). *Environmental marketing management: Meeting the green challenge*. Financial Times. Pitman Publishing.

Peattie, K. (2001). Delivered by publishing technology to: Chinese University of Hong Kong delivered by publishing technology to : Chinese University of Hong Kong. *The Marketing Review, 2*, 129–146.

Penrose, E. T. (1959). *The theory of the growth of the firm*. Basil Blackwell.

Perrini, F., & Tencati, A. (2006). Sustainability and stakeholder management: The need for new corporate performance evaluation and reporting systems. *Business Strategy and the Environment, 15*, 296–308.

Pesce, M., Tamai, I., Guo, D., Critto, A., Brombal, D., Wang, X., Cheng, H., & Marcomini, A. (2020). Circular economy in China: Translating principles into practice. *Sustainability (Switzerland), 12*(3), 1–31. https://doi.org/10.3390/su12030832

Porter, M. E., & Kramer, M. R. (2003). The competitive advantage of corporate philanthropy. In *Harvard Business Review on Corporate Responsibility*. Harvard Business School Press.

Porter, M. E., & Kramer, M. R. (2011). Creating shared value. *Harvard Business Review*, 63–77. https://doi.org/10.1108/09600039410055963

Porter, M. E., van der Linde, C., & van der Linde, C. (1995). Toward a new conception of the environment-competitiveness relationship. *Journal of Economic Perspectives, 9*(4), 97–118. https://doi.org/10.1257/jep.9.4.97

Prakash, A. (2000). Responsible Care: An assessment. *Business and Society, 39*(2), 183–209.

Preston, F. (2012). *Briefing paper a global redesign? Shaping the circular economy*. Chatham House.

Preuss, L. (2009). Addressing sustainable development through public procurement: The case of local government. *Supply Chain Management: An International Journal, 14*(3), 213–223. https://doi.org/10.1108/13598540910954557

Radin, M. (1932). The endless problem of corporate personality. *Columbia Law Review, 32*, 643–667.

Rao, P., & Holt, D. (2005). Do green supply chains lead to competitiveness and economic performance? *International Journal of Operations and Production Management, 25*(9), 898–916. https://doi.org/10.1108/01443570510613956

Reijnders, L. (1997). The factor X debate: Setting targets for eco-efficiency. *Journal of Industrial Ecology, 2*(1).

Reinhardt, F. L. (2000). Sustainability and the firm. *Interfaces, 30*(3), 26–41.

Rex, E., & Baumann, H. (2007). Beyond ecolabels: What green marketing can learn from conventional marketing. *Journal of Cleaner Production, 15*(6), 567–576.

Robèrt, K.-H. (2000). Tools and concepts for sustainable development, how do they relate to a general framework for sustainable development, and to each other? *Journal of Cleaner Production, 8*(3), 243–254. https://doi.org/10.1016/S0959-6526(00)00011-1

Robèrt, K.-H., Schmidt-Bleek, B., Aloisi de Larderel, J., Basile, G., Jansen, J. L., Kuehr, R., Price Thomas, P., Suzuki, M., Hawken, P., & Wackernagel, M. (2002). Strategic sustainable development - selection, design and synergies of applied tools. *Journal of Cleaner Production, 10*, 197–214.

Roca, L. C., & Searcy, C. (2012). An analysis of indicators disclosed in corporate sustainability reports. *Journal of Cleaner Production, 20*(1), 103–118. https://doi.org/10.1016/j.jclepro.2011.08.002

Roome, N., & Louche, C. (2016). Journeying toward business models for sustainability: A conceptual model found inside the black box of organisational transformation. *Organization & Environment, 29*(1), 11–35. https://doi.org/10.1177/1086026615595084

Russo, M. V., & Fouts, P. A. (1997). A resource-based perspective on corporate environmental performance and profitability. *Academy of Management Journal, 40*(3), 534–559. https://doi.org/10.2307/257052

SAI. (2007). Overview of SA 8000. http://www.sa-intl.org/index.cfm?fuseaction=Page.viewPage&pageId=473

Sarkis, J., & Cordeiro, J. J. (2001). An empirical evaluation of environmental performance: Pollution prevention versus end-of-pipe. *European Journal of Operational Research, 135*(1), 102–113. https://doi.org/10.1016/S0377-2217(00)00306-4

Schaltegger, S., Hansen, E. G., & Lüdeke-Freund, F. (2016). Business models for sustainability: Origins, present research, and future avenues. *Organization & Environment, 29*(1), 3–10. https://doi.org/10.1177/1086026615599806

Schaltegger, S., & Wagner, M. (2006). Integrative management of sustainability performance, measurement and reporting. *International Journal of Accounting, Auditing and Performance Evaluation, 3*(1), 1–19. https://doi.org/10.1504/IJAAPE.2006.010098

Schwager, P., & Moser, F. (2005). The application of chemical leasing business models in Mexico. *Environ Sci & Pollut Res, 13*(2), 131–137. https://doi.org/10.1065/espr2006.02.294

Schwartz, B., & Tilling, K. (2009). 'ISO-lating' corporate social responsibility in the organizational context: A dissenting interpretation of ISO 26000. *Corporate Social Responsibility and Environmental Management, 16*(5), 289–299. https://doi.org/10.1002/csr.211

Seppälä, J., Koskela, S., Melanen, M., & Palperi, M. (2002). The Finnish metals industry and the environment. *Resources, Conservation and Recycling, 35*(1–2), 61–76. https://doi.org/10.1016/S0921-3449(01)00122-7

Seuring, S., & Müller, M. (2008a). Core issues in sustainable supply chain management. *Business Strategy and the Environment, 17*, 455–466.

Seuring, S., & Müller, M. (2008b). From a literature review to a conceptual framework for sustainable supply chain management. *Journal of Cleaner Production, 16*(15), 1699–1710. https://doi.org/10.1016/j.jclepro.2008.04.020

Sheth, J. N., Parvatiyar, A., Sharma, A., & Sheth, J. N. (1995). The evolution of relationship marketing: A framework for analysis. *International Business Review, 4*(4), 397–418.

Siebenhüner, B., & Arnold, M. (2007). Organizational learning to manage sustainable development. *Business Strategy and the Environment, 16*, 339–353.

Skouloudis, A., & Evangelinos, K. (2009). Sustainability reporting in Greece: Are we there yet? Enviornmental Quality Management, Autumn, 43–60. https://doi.org/10.1002/tqem.

Smith, C. (2003). The new corporate philanthropy. In *Harvard Business Review on Corporate Responsibility*. Harvard Business School Press.

Stindt, D. (2017). A generic planning approach for sustainable supply chain management - how to integrate concepts and methods to address the issues of sustainability? *Journal of Cleaner Production, 153*, 146–163. https://doi.org/10.1016/j.jclepro.2017.03.126

Stubbs, W., & Cocklin, C. (2008). Conceptualizing a "Sustainability Business Model". *Organization & Environment, 21*(2), 103–127. https://doi.org/10.1177/1086026608318042

Swift, T., & Zadek, S. (2002). *Corporate responsibility and the competitive advantage of nations*. The Copenhagen Centre Accountability.

Szekely, F., & Knirsch, M. (2005). Responsible leadership and corporate social responsibility: Metrics for sustainable performance. *European Management Journal, 23*(6), 628–647.

The Economist. (2005). *The good company. A survey of corporate social responsibility*. The Economist.

The Ellen MacArthur Foundation. (2015). *Towards a circular economy: Business rationale for an accelerated transition*. Ellen MacArthur Foundation (EMF).

UN. (2002). *World summit on sustainable Development plan of implementation*. United Nations. http://www.johannesburgsummit.org/html/documents/summit_docs/0409_plan_final.pdf

UNEP. (1990). Resource efficient and cleaner production. http://www.unep.fr/scp/cp/

UNEP. (2001). Cleaner production (CP) activities. United Nations Environemt Programme. http://www.uneptie.org/PC/cp/home.htm

UNEP, & SETAC. (2007). Life cycle management - a business guide to sustainability.

UNGC. (2008). The ten principles. United Nations Global Compact. http://www.unglobalcompact.org/AboutTheGC/TheTenPrinciples/index.html

UNIDO. (2011). Global promotion and implementation of chemical leasing business models in industry. Progress Report 2011. http://chemicalleasing.com/docs/ChL_annualreport2011_final.pdf

UNIDO, Raynard, P., & Forstater, M. (2002). Corporate social responsibility. In *Implications for Small and Medium Enterprises in Developing Countries*. United Nations Industrial Development Organization (UNIDO).

United Nations Environment Programme, & UNEP. (2006). *The Strategic Approach to International Chemicals Management – full text and resolutions of the International Conference on Chemical Management 2003*. http://www.saicm.org/images/saicm_documents/saicm.texts/SAICM_publication_ENG.pdf.

UNU. (2007). *Factor X*. United Nations UniversityNTTIwate Prefectural Government. http://www.ias.unu.edu/ecology/g_economy/factorx.htm

Upward, A., & Jones, P. (2016). An ontology for strongly sustainable business models: Defining an Enterprise framework compatible with natural and social science. *Organization and Environment, 29*(1), 97–123. https://doi.org/10.1177/1086026615592933

Urbinati, A., Chiaroni, D., & Chiesa, V. (2017). Towards a new taxonomy of circular economy business models. *Journal of Cleaner Production, 168*, 487–498. https://doi.org/10.1016/j.jclepro.2017.09.047

van Buren, N., Demmers, M., van der Heijden, R., & Witlox, F. (2016). Towards a circular economy: The role of Dutch logistics industries and governments. *Sustainability (Switzerland), 8*(7), 1–17. https://doi.org/10.3390/su8070647

van Marrewijk, M., & Hardjono, T. W. (2003). European corporate sustainability framework for managing complexity and corporate transformation. *Journal of Business Ethics, 44*(2/3), 121.

Verdeyen, V., Put, J., & van Buggenhout, B. (2004). A social stakeholder model. *International Journal of Social Welfare, 13*, 325–331.

von Weizsäcker, E., Hargroves, K., Smith, M. H., Desha, C., & Stasinopoulos, P. (2009). Factor five. Transforming the global economy through 80% improvements in resource productivity. *Earth*.

von Weizsäcker, E., Lovins, A. B., & Lovins, L. H. (1998). Factor four. Doubling wealth, halving resource use. *Earth*.

Wai, A., Cheung, K., & Cheung, W. K. (2017). Do stock investors value corporate sustainability ? *Evidence from an Event Study., 99*(2), 145–165.

Warhurst, A. (2002). Sustainability indicators and sustainability performance management. In *Mining, minerals and sustainable Development* (Vol. 43). International Institute for Environment and DevelopmentWorld Business Council for Sustainable Development.

WBCSD. (2000). Eco-efficiency: Creating more value with less impact.

WBCSD. (2002). *Corporate social responsibility. The WBCSD's journey*. World Business Council for Sustainable Development.

WBCSD. (2004). *A sustainable livelihoods approach to industry challenges: The south African sugar industry*. World Business Council for Sustainable Development. http://www.wbcsd.org/plugins/docsearch/details.asp?txtDocTitle=sustainablelivelihoods&DocTypeId=24&CharValList=24;&ObjectId=NzQ1OQ&URLBack=result.asp%253FtxtDocTitle%253Dsustainable+livelihoods%2526DocTypeId%253D24%2526CharValList%253D24%253B%2526SortOrder%253D%2526CurPage%253

WBCSD, & UNEP. (1998). *Cleaner production and eco-efficiency complementary approaches to sustainable Development* (p. 12). WBCSD, & UNEP.

Webster, K. (2013). What might we say about a circular economy? Some temptations to avoid if possible. *World Future: The Journal of New Paradigm Research, 69*(7–8), 542–554. https://doi.org/10.1080/02604027.2013.835977

Welford, R. (2005). Corporate social responsibility in Europe, North America and Asia. 2004 survey results. *Journal of Corporate Citizenship, 17*, 33–52.

Weymes, E. (2004). Management theory. Balancing individual freedom with organisational needs. *Journal of Corporate Citizenship, 16*, 85–98.

Wilenius, M. (2005). Towards the age of corporate responsibility? Emerging challenges for the business world. *Futures, 37*, 133–150.

Willard, B. (2002a). *The sustainability advantage*. New Society Publishers.

Willard, B. (2002b). *The sustainability advantage*. New Society Publishers.

Yong, R. (2007). The circular economy in China. *Journal of Material Cycles and Waste Management, 9*(2), 121–129. https://doi.org/10.1007/s10163-007-0183-z

Yuan, Z., Bi, J., & Moriguichi, Y. (2008). The circular economy: A new Development strategy in China. *Journal of Industrial Ecology, 10*(1–2), 4–8. https://doi.org/10.1162/108819806775545321

Zadek, S. (1999). Stalking sustainability. *GMI, 26*, 1–11.

Chapter 5
Organisational Change Management for Sustainability

5.1 Introduction

Organisational change aims to move from the current state to a more desirable state (Ragsdell, 2000), ranging from minor to radical changes (Dawson, 1994). Change represents an opportunity; it must be anticipated, prepared for, and managed (European Commission, 1998). Organisations that refuse to change run the risk of being sidelined by external events (Collins & Porras, 2002; Lozano & Barreiro-Gen, 2021), for example, changes in government regulations, technologies, products, workforce, competition, or epidemics. Such external events force companies to make moderate organisational changes at least once a year and major changes every 4 or 5 years (Kotter & Schlesinger, 1979; Lozano & Barreiro-Gen, 2021).

The rate of change is dependent on the organisations context, nature, and external events (Lozano & Barreiro-Gen, 2021). Changes in organisations tend to be complex (Dawson, 1994), continuous, iterative, and uncertain (Pettigrew & Whipp, 1991). It is fairly easy to identify organisational changes once they have occurred, more difficult to analyse them while they are ongoing, and even more difficult to predict their direction and tempo (Benne & Birnbaum, 1969).

Major changes take time; therefore, a sense of urgency and importance needs to be attached to the starting points of successful change processes (Kotter, 1996). Any change will impinge on vested interests and violate territorial rights (Shephard, 1969), disrupting and unsettling status (Maurer, 1996), where initiating a new order of things can be very difficult or dangerous to handle (Machiavelli, 1966).

Organisational changes need to go beyond mere changes in technology or management systems, since they require changes in culture (Cannon, 1994; Doppelt, 2003; Lozano, 2013) and in the system elements (Lozano, 2020).

R. Lozano, *Toward Sustainable Organisations*, Strategies for Sustainability,
https://doi.org/10.1007/978-3-030-99676-5_5

5.2 Change Typologies

Four categorisations can be helpful in understanding change, including (1) stakeholder focus, (2) change pathways, (3) organisational focus, and (4) intervention.

Traditionally, there are two types of change, according to **stakeholder focus** (Freeman, 1984; Lozano, 2013): (1) *proactive* (or internal), which is focused on constantly reassessing objectives and policies that affect or are affected by primary stakeholders. Organisations have a higher degree of control over these types of change. (2) *reactive* (or externally driven), referring to change that happens on the outskirts of the company's sphere of influence, i.e. secondary stakeholders, and can take different forms (e.g. political, economic, or social).

Three main **change pathways** to facilitate change have been proposed: (1) *developmental*, (2) *transitional* ("three-step strategy"), and (3) *transformational.*

Developmental change represents the improvement of an existing skill, method, performance standard, or condition (Anderson & Ackerman Anderson, 2001). It is the simplest of the three types of change, where the new state is a prescribed enhancement of the old one, rather than a radical change.

The transitional approach, based on the seminal work of Lewin, 1947a, 1947b, proceeds via three steps: (1) unfreezing the present level or situation, (2) moving to the new level, and (3) freezing at the new level. Transitional change is more complex since; instead of improving the existing state, it is replaced with an entirely different one (Anderson & Ackerman Anderson, 2001). In transitional change, the first step is to break away from the *status quo* (SQ), i.e. current paradigms, mental models, routines, and practices, in order to enter a transition period, and finally a new state, or status quo novo (SQN). A modification of the transitional change proposes a framework, with three stages (conception, transition, and operation) that are affected by the substance, politics, and context of the change (Dawson, 1994).

Some authors have improved on Lewin's strategy, such as Dawson (1994), who proposes a three step approach, i.e. conception, transition, and operation, affected by the substance, politics, and context of the change.

Transformational change is based on a radical shift from one state to another, such that a new order arises out of existing crisis or chaos (Anderson & Ackerman Anderson, 2001). It is more complex than the previous.

According to the **organisational focus**, four types of change have been proposed (Lorenzi & Riley, 2000), which overlap in some cases: (1) *operational* changes, affecting the way the ongoing operations of the business are conducted; (2) *strategic* changes, impacting business direction; (3) *cultural* changes, affecting the basic organisational philosophies by which the business is conducted; and (4) *political* changes in staffing, occurring primarily for political reasons of various types.

According to the **intervention approach**, three types of change have been proposed (Lozano, 2013) and (Bennis et al., 1969): (1) *non-intervention*, which relies on serendipitous change, where there is little or no direction or guidance in accordance with the laissez-faire doctrine of the "invisible-hand"; (2) *radical intervention*, emphasising conflict and class struggle, which may restrict the freedom of individuals or groups; and (3) *planned change*, which offers the advantage of some type of guidance, without being too constrictive or too serendipitous, but rests on collaboration, criticised values, evaluated practical experience, and research knowledge.

Of the intervention types, planned changes are the most flexible (Bennis et al., 1969; Lozano, 2013). These entail (1) identification of mission and values, (2) collaboration and conflict, (3) control and leadership, (4) resistance and adaptation to change, (5) utilisation of human resources, (6) communication, and (7) management development (Bennis et al., 1969). Planned change needs to consider the content, context, and process of change (Pettigrew & Whipp, 1991). The effectiveness of planned change is often related to the participation of members, at all levels of an organisation, in assessing and diagnosing needful change and in formulating its goals and objectives (Benne & Birnbaum, 1969).

Planned organisational change (see Fig. 5.1) is aimed at taking the organisation from the *status quo* (SQ) to a new stage or equilibrium point, *status quo novo* (SQN) (Lozano, 2013), equivalent to Lewin's, 1947a model and Anderson and Ackerman Anderson (2001) transitional changes. In this process the system's inertia needs to be temporarily disrupted by applying leverage in the right places. This brings the system to a temporary transition period, where the changed balance of forces adjusts to each other and reach the SQN. Once the balance of forces is adjusted, the new structure and goals are set, and the SQN then starts becoming the next SQN. This process is shown in Fig. 5.1. If the system is unstable, the leverage will upset it, and the transition period will become perennial and thus the SQN would not be reached.

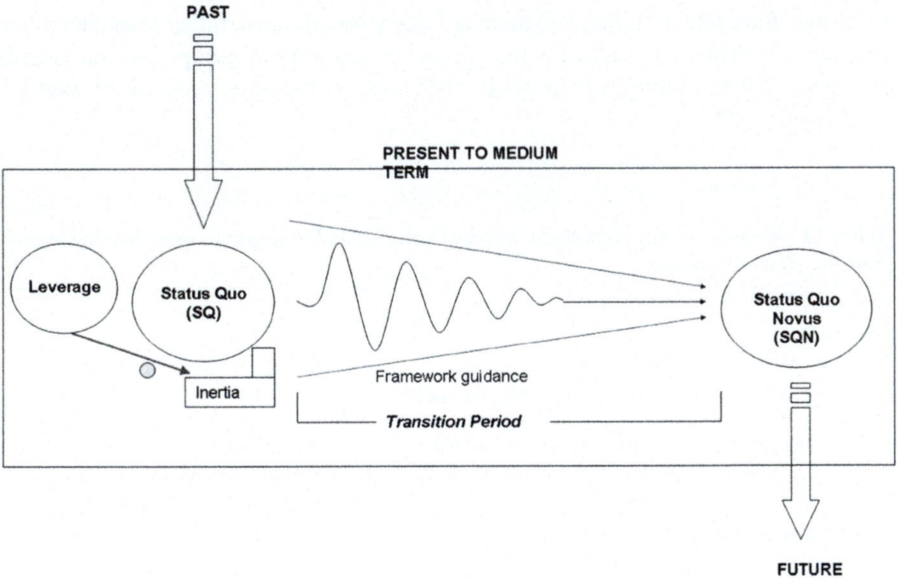

Fig. 5.1 Organisational change, moving from the status quo (SQ) to the status quo novo (SQN). Source: (Lozano, 2013)

5.3 Change Drivers and Resistance to Change

Change drivers are the leverage that disrupt the organisational system SQ and push change (Lozano, 2015; Lozano & von Haartman, 2018; Maurer, 1996; Senge, 1999). Although organisations are heterogeneous, the most commonly recognised change drivers include:

- Proactive leadership, or management (Dawson, 1994).
- Economic benefits, or a failure to obtain them (Cannon, 1994).
- Fear, which produces extraordinary short-term changes, but with negative vision and aspiration, which produces continuous learning and growth, and has positive vision (Senge, 1999).
- External factors, such as political and financial upheaval, new technologies, regulatory change, worldwide competition, and consumer preferences (Dawson, 1994; European Commission, 1998).
- Diagnosis of something being wrong in the organisation and needing to be changed (Carr, 2001).
- The upsurge of visible crises that can attract attention and push up urgency levels (Kotter, 1996).

Organisational changes that threaten the SQ are affected by resistance forces (Bovey & Hede, 2001; Lozano, 2013; Maurer, 1996; Senge, 1999), where the larger the change, the stronger the resistance (Maurer, 1996). Resistance can be covert or explicit with blatant struggles over resources, expressions of doubt, and unwillingness to commit to the change efforts (Lewin, 1947b). Resistance is easier to attend to it when the cycle is young, and there are many possibilities, rather than later when resistance is firmly entrenched, which relates to maturity of groups and individuals (Lozano, 2013). Table 5.1 presents the most commonly recognised barriers to change.

Table 5.1 Most commonly recognised barriers to change and strategies to overcome the barriers from the literature review

System level		
Individual	Group	Organisational
• Misunderstanding/lack of communication • Lack of trust • Threat to job status/security	• Group culture • Ignoring institutions in the group • Individual-group conflict	• Lack of strategy/long term plans (managerial) • Bureaucracy/patriarchal models (organisational) • Lack of top management commitment/ walking the talk (managerial)

Source: Lozano (2013)

5.4 Change Incorporation and Institutionalisation

Incorporation refers to the initiation of change (see Lozano, 2006), i.e. the first stage once the change drivers have acted upon the organisation's system that take the organisation into a transition period (Lozano, 2013). Doppelt (2003) proposed seven initiation, or incorporation, stages:

1. Change the dominant mindset, out of which the current system arose.
2. Rearrange the parts of the system.
3. Alter the goals of the system.
4. Restructure the rules of engagement of the system.
5. Shift the flow of information and communication in the system.
6. Correct the feedback loops of the system.
7. Adjust the parameters of the system.

It should be noted that if change is not made part of the culture, norms, and values of an organisation, it can be undone, even years after the incorporation started (Kotter, 1996). For example, as much as three-quarters of the most popular approaches to organisational change, e.g. total quality management (TQM), i.e. strategic planning, re-engineering, and downsizing, achieve no success because they fail to change underlying patterns and employees' behaviours; while modifying the structure, thinking and perspectives of the system could be clearer (Senge, 1999). Such approaches manage organisational changes from a technical viewpoint, neglecting the human element, which leads to high levels of resistance (Bovey & Hede, 2001).

When change is adopted and put into practice for long enough, and increasingly, by different members of the institution, until widespread implementation and stabilisation is achieved, they become part of the culture (Rogers, 1962; van de Ven et al., 1999), i.e. change becomes institutionalised. Institutionalisation, thus, is the maintenance of the new stable state (Lewin's (1947a) new level or Anderson and Ackerman Anderson (2001) new state).

Incorporation and institutionalisation provide a more dynamic perspective on change pathways, as illustrated in Fig. 5.2. This results in alternated transition and institutionalisation periods.

5.5 Organisational Change Management for Sustainability

A limited number of organisations have successfully incorporated sustainability into their systems and culture (Hussey et al., 2001; Linnenluecke & Griffiths, 2010; Siebenhüner & Arnold, 2007). Many sustainability approaches have been based on technocentric solutions and managerial ploys, which neglect culture, the supply chain, and the interactions between the system elements and the four dimensions of sustainability (Lozano, 2012c, 2015).

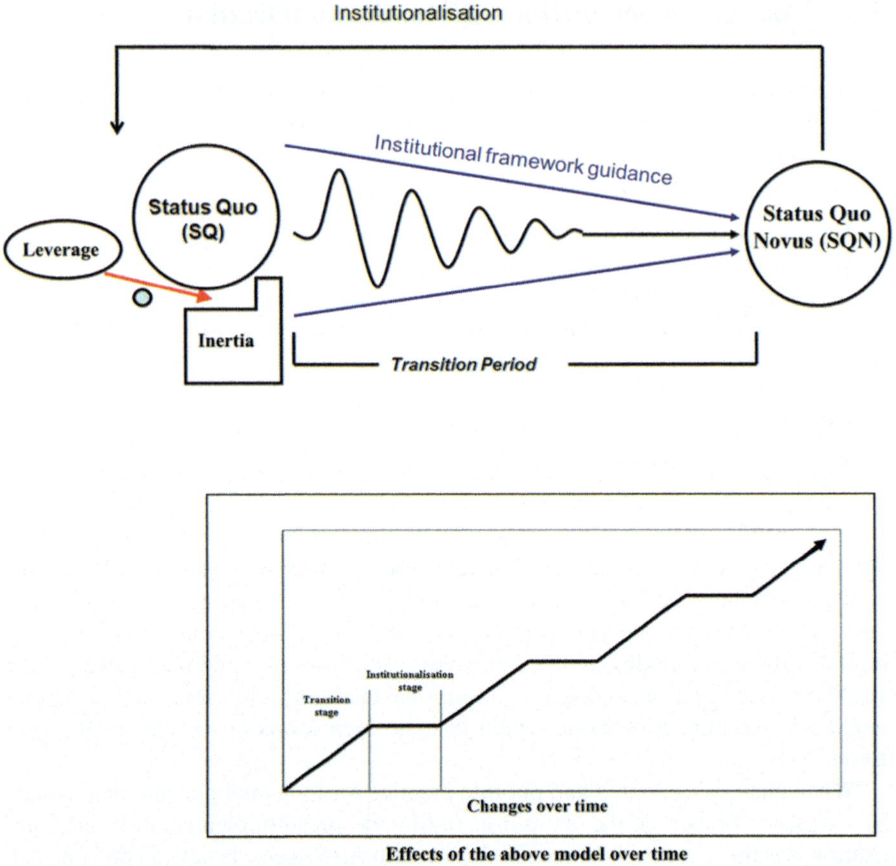

Fig. 5.2 Combining developmental and transitional changes to create transformational changes

Change management provides a dynamic and systemic perspective to organisational sustainability (Lozano, 2013), by aiming to move from the current state to one that is more desirable (Ragsdell, 2000), ranging from minor to radical changes (Dawson, 1994, 2001; Rogers, 1962). Long-lasting sustainability change requires a holistic perspective (Hjorth & Bagheri, 2006; Linnenluecke et al., 2009; Lozano & Huisingh, 2011).

Organisational change management focuses on "soft issues" (values, visions, philosophies, policies, and employee empowerment), i.e. change management practices (Doppelt, 2003; Dunphy et al., 2003), organisational and behavioural change management practices (Lozano, 2012b), drivers for sustainability (Lozano, 2015; Lozano & von Haartman, 2018; Sayce et al., 2007), barriers to change (Blanco-Portela et al., 2017; Lozano, 2013), and the link between organisational change management and assessment and reporting, with, in many cases, reinforcing effects (Ceulemans et al., 2015; Domingues et al., 2017; Lozano et al., 2016). Sustainability assessment and reporting have been considered to be an important catalyst for sustainability change (Adams & McNicholas, 2007; Doppelt, 2003; Lozano, 2015; Lozano et al., 2016).

Sustainability in organisations has been, in general, equally driven by external stimuli and internal factors and then followed mainly by internal factors but with some external stimuli (Lozano, 2018). Internally, changes can be done through managerial measurement and control (Henriques & Richardson, 2005) or by focusing on internal change and innovation (Doppelt, 2003; Henriques & Richardson, 2005). The former rely on strategic top-down changes, while the latter consist of participative collaborative changes (Lozano, 2013). The organisations that have engaged in sustainability have done so mainly through upper management level initiatives (Siebenhüner & Arnold, 2007).

Lozano (2012a, 2013, 2015), based on the works of Bennis et al. (1969), Lewin (1947a, 1947b), Anderson and Ackerman Anderson (2001), and Luthans (2002), proposed an "Orchestrating Change for Corporate Sustainability' framework (see Fig. 5.3) to help explain the process of organisational changes for corporate sustainability. The framework posits that "Orchestrated planned change can disrupt the status quo (SQ) and help move towards a more sustainability-orientated state (MSOS), in a continuously iterative process. The entire system and its elements, as well as their respective attitudes, need to be addressed. In this process, it is important to foster the drivers to change, and apply the appropriate strategies to overcome the barriers to change. The institutional framework can help to maintain stability during the changes, and thus facilitate sustainability institutionalisation. During these changes, the system would pass through a transitional period, where the different balance of forces adjust to each other, to reach the MSOS. Once all the forces are rebalanced, and the new structure and goals are set, the MSOS starts becoming the status quo novo (SQN). Because of the dynamism of sustainability, the process has to start again after stabilisation. Planning organisational changes, whilst engaging with the different organisational levels and their attitudes, could help companies to better overcome resistance to change and integrate their efforts for sustainability more holistically, including technological and human changes, i.e. taking a more holistic perspective to contribute to sustainability".

5.5.1 Sustainability Change Drivers and Resistance to Change

Sustainability has been driven by many factors (Hopkins, 2002; Oskarsson & von Malmborg, 2005; Salzmann et al., 2005). Traditionally, these have been divided into external and internal drivers.

External drivers include national policies (e.g. the French stock exchange to report on sustainability (MacLeod & Lewis, 2004) and the EU directive on corporate sustainability reporting (European Parliament, 2014)); social action under administrative guidance (gyosei-shido), imperatives in Japanese society, business

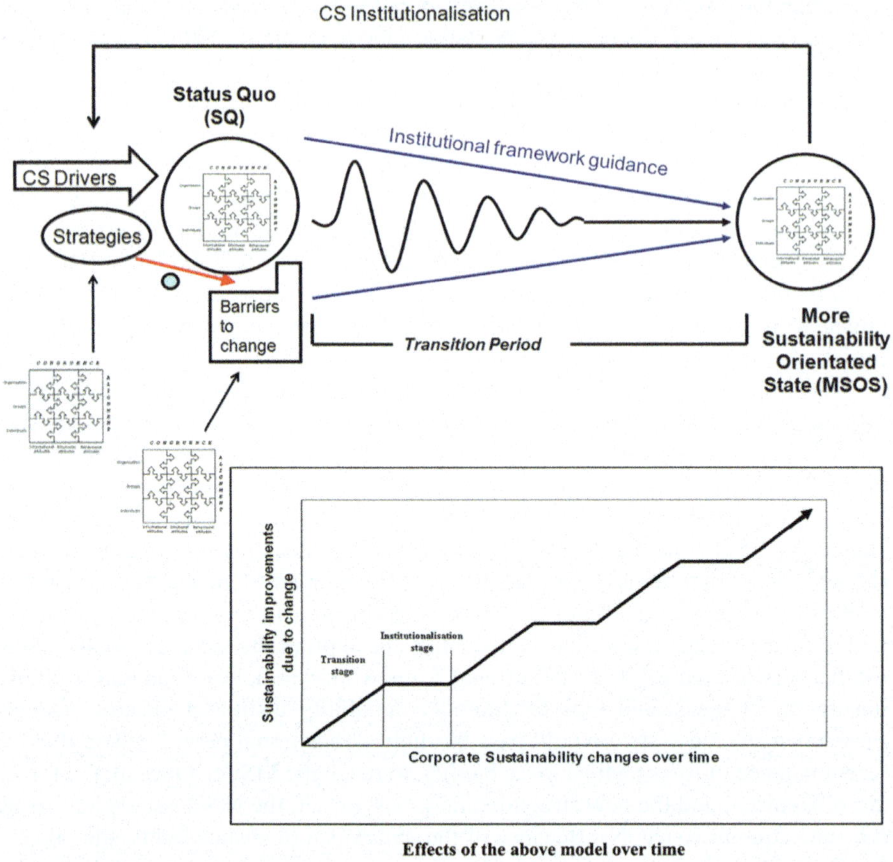

Fig. 5.3 Framework for explaining the dynamics of Orchestrating Change for Corporate Sustainability. Source (Lozano, 2013)

leadership, government, and universities (Fukukawa & Moon, 2004); and NGOs, including stakeholder pressure (Fernández et al., 2006; Frehs, 2003; Zadek, 1999).

Internal drivers include ethical leadership (recognised to be one of the key elements for the successful introduction, implementation and institutionalisation of change) (Dawson, 1994; DeSimone & Popoff, 2000; Doppelt, 2003; Lozano, 2015), risk management and protection of business reputation (Ditlev-Simonen & Atle, 2011; Lantos, 2001), improvements in economic values (C.E.C., 2001; Carroll, 1999; Lantos, 2001), and enhancements in corporate image (Frehs, 2003).

Lozano (2015) complemented the binary (external/internal) perspective by adding a connecting dimension and compiling a list of internal, connecting, and external drivers, as presented in Fig. 5.4. An analysis of driver importance shows five ranking groups (see Fig. 5.5): (1) highest, including proactive leadership and reputation; (2) high, including moral and ethical obligations, increased levels of social awareness, organisation culture, and regulation and legislation; (3) medium,

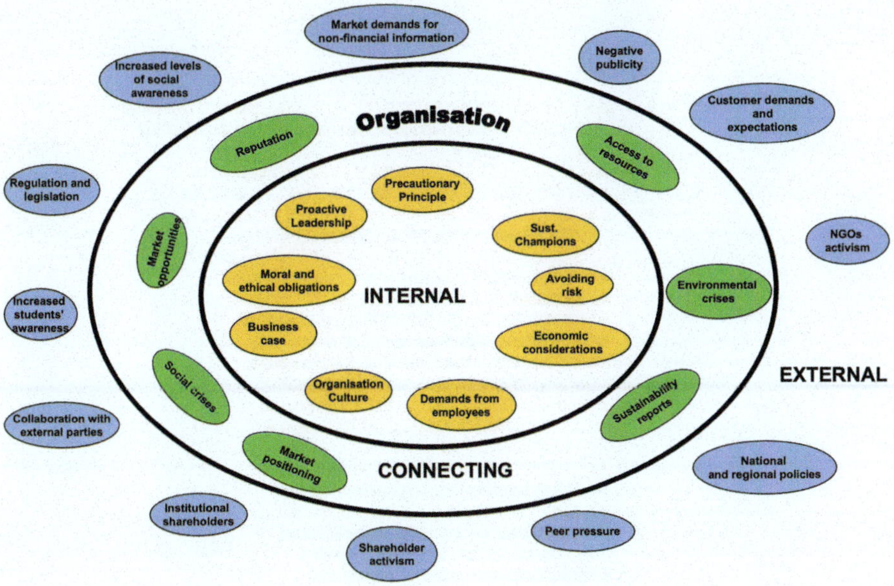

Fig. 5.4 Corporate sustainability drivers (Lozano, 2015). Yellow indicates internal drivers, green the connecting drivers, and blue the external drivers

including collaboration with external parties, avoiding risk, national and regional policies, sustainability reports, customer demands and expectations, market positioning, business case, and precautionary principle; (4) low, including economic considerations, market opportunities, sustainability champions, environmental crises, increased students' awareness, and access to resources; and (5) lowest, including social crises, demands from employees, market demands for non-financial information, shareholder activism, NGOs activism, negative publicity, peer pressure, and institutional shareholders.

Sustainability change resistance is manifested through barriers to change (Lozano, 2013), such as (1) external, including pressure from competitors (Magretta, 2000; Rosner, 1995), regulation and legislation (C.E.C., 2002; DeSimone & Popoff, 2000; Magretta, 2000), lack of knowledge about impacts on and from suppliers and customers (DeSimone & Popoff, 2000; Rosner, 1995), and lack of interest from consumers or investors (C.E.C., 2002) and (2) internal (see Lozano, 2013), including unwillingness to change, narrow focus of sustainability, lack of systems thinking, faith in technological solutions, lack of management commitment, failure to institutionalise sustainability, i.e. not changing the cultural and mental models, organisational structures inhibiting collaboration, and unsuccessful incorporation attempts. In general, organisation can address internal resistance easier than external ones (Lozano, 2012a, 2013).

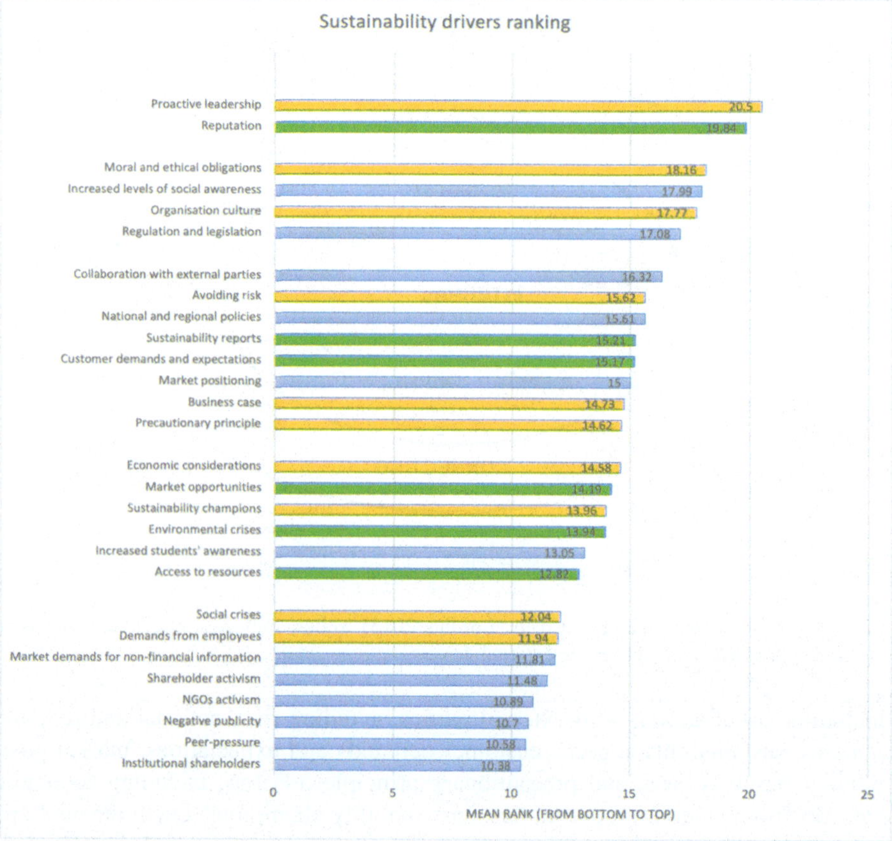

Fig. 5.5 Rankings of sustainability drivers for organisations. Source: (Lozano & von Haartman, 2018)

5.5.2 *Sustainability Incorporation and Institutionalisation in Organisations*

Limited research has been carried out on the incorporation and institutionalisation of sustainability in organisations. Firstly, a study on sustainability incorporation discussed as a process of diffusion of innovations and institutionalisation at its adoption and implementation in the context of higher education institutions (Lozano, 2006). Secondly, empirical research through a survey answered by 281 organisations (from civil society, corporations, and public sector) focusing on system element changes, from the start (i.e. incorporation) and affected by the changes (i.e. institutionalisation), found that the main system elements have been governance, management and strategy, and operations and production and that the majority of the changes were effected between six and seven systems elements (indicating a large degree of institutionalisation) (Lozano & Garcia, 2020).

5.6 Discussion

During the two last decades, there has been an increase in research on organisational change management for sustainability. Organisational change aims to move from the current state to one that is more desirable, ranging from minor to radical changes. Change represents an opportunity that must be anticipated, prepared for, and managed. Changes in organisations are complex, continuous, iterative, and uncertain. The rate of change depends on the organisation's context, nature, and external events. Organisational change management provides a dynamic perspective through addressing the time dimension of sustainability. It aims to move from the current state to one that is more desirable.

Change management for sustainability must encompass all the system elements in a process where the status quo is disrupted and a new, more sustainability-oriented state is achieved. In this process, the drivers to change must be fostered, and resistance to change must be overcome, in order to institutionalise the changes. The process is iterative, multilevelled, and highly dynamic. Many organisations are proactively engaging with sustainability, where sustainability has been, in general, equally driven by external stimuli and internal factors, followed mainly by internal factors, but with some external stimuli.

Planning sustainability changes must address its four dimensions holistically and include the technical, managerial, and organisational issues, not forgetting the organisation's stakeholders.

References

Adams, C. A., & McNicholas, P. (2007). Making a difference. Sustainability reporting, accountability and organisational change. *Accounting, Auditing & Accountability Journal, 20*(3), 382–402. https://doi.org/10.1108/09513570710748553

Anderson, D., & Ackerman Anderson, L. (Eds.). (2001). *Beyond change management.* Jossey-Bass/Pfeiffer (Wiley).

Benne, K. D., & Birnbaum, M. (1969). Principles of changing. In W. G. Bennis, K. D. Benne, & R. Chin (Eds.), *The planning of change* (Second). Holt, Rinehart and Winston, Inc.

Bennis, W. G., Benne, K. D., & Chin, R. (1969). *The planning of change* (Second). Holt, Rinehart and Winston, Inc.

Blanco-Portela, N., Benayas, J., Pertierra, L. R. L. R., & Lozano, R. (2017). Towards the integration of sustainability in higher education institutions: A review of drivers of and barriers to organisational change and comparison against those found of companies. *Journal of Cleaner Production, 166*, 563–578. https://doi.org/10.1016/j.jclepro.2017.07.252

Bovey, W. H., & Hede, A. (2001). Resistance to organisational change: The role of defence mechanisms. *Journal of Managerial Psychology, 16*(7), 534–548.

Carroll, A. B. (1999). Corporate social responsibility: Evolution of a definitional construct. *Business Society, 38*(3), 268–295.

C.E.C. (2001). *Promoting a European framework for corporate social responsibility.* Commission of the European Communities.

C.E.C. (2002). *Corporate social responsibility: A business contribution to sustainable development.* Commission of the European Communities.

Cannon, T. (1994). *Corporate responsibility. A textbook on business ethics, governance, environment: Roles and responsibilities*. Pitman publishing.

Carr, A. (2001). Understanding emotion and emotionality in a process of change. *Journal of Organizational Change, 14*(5), 421–434.

Ceulemans, K., Lozano, R., & Alonso-Almeida, M. (2015). Sustainability reporting in higher education: Interconnecting the reporting process and Organisational change Management for Sustainability. *Sustainability, 7*(7), 8881–8903. https://doi.org/10.3390/su7078881

Collins, J., & Porras, J. I. (2002). *Built to last. Successful habits of visionary companies*. Harper Business Essentials.

Dawson, P. (1994). *Organizational change. A processual approach*. Paul Chapman Publishing Ltd.

Dawson, P. (2001). *Organizational change. International encyclopedia of the social & behavioral sciences* (pp. 10921–20923). Elsevier.

DeSimone, L. D., & Popoff, F. (2000). *Eco-efficiency. The business link to sustainable development*. MIT Press.

Ditlev-Simonen, C. D., & Atle, M. (2011). What motivates managers to pursue corporate responsibility? A survey among key stakeholders. *Corporate Social Responsibility and Environmental Management, 18*(1), 25–38. https://doi.org/10.1002/csr.237

Domingues, A. R., Lozano, R., Ceulemans, K., & Ramos, T. B. (2017). Sustainability reporting in public sector organisations: Exploring the relation between the reporting process and organisational change management for sustainability. *Journal of Environmental Management, 192*, 292–301. https://doi.org/10.1016/j.jenvman.2017.01.074

Doppelt, B. (2003). *Leading change toward sustainability. A change-management guide for business, government and civil society*. Greenleaf Publishing.

Dunphy, D., Griffiths, A., & Benn, S. (2003). *Organizational change for corporate sustainability (Third)*. Routledge.

European Commission. (1998). *Managing change. European Commission*. Employment & Social Affairs.

European Parliament. (2014). Directive 2014/95/EU of the European Parliament and of the council of 22 October 2014 amending directive 2013/34/EU as regards disclosure of non-financial and diversity information by certain large undertakings and groups.

Fernández, E., Junquera, B., & Ordiz, M. (2006). Managers' profile in environmental strategy: A review of the literature. *Corporate Social Responsibility and Environmental Management, 13*, 261–274. https://doi.org/10.1002/csr.109

Freeman, R. E. (1984). *Strategic management. A stakeholder approach*. Pitman Publishing Inc.

Frehs, J. (2003). *Corporate social responsibility: Lessons learned*. Natural Resources Canada.

Fukukawa, K., & Moon, J. (2004). A Japanese model of corporate social responsibility? *Journal of Corporate Citizenship, 16*, 45–59.

Henriques, A., & Richardson, J. (2005). *The triple bottom line. Does it all add up?* Earthscan.

Hjorth, P., & Bagheri, A. (2006). Navigating towards sustainable development: A system dynamics approach. *Futures, 38*(1), 74–92. https://doi.org/10.1016/j.futures.2005.04.005

Hopkins, M. J. D. (2002). Sustainability in the internal operations of companies. *Corporate Environmental Strategy, 9*(2), 1–11.

Hussey, D. M., Kirsop, P. L., & Meissen, R. E. (2001). Global reporting initiative guidelines: An evaluation of sustainable development metrics for industry. *Environmental Quality Management*, 1–20.

Kotter, J. P. (1996). *Leading change*. Harvard Business School Press.

Kotter, J. P., & Schlesinger, L. A. (1979). Choosing strategies for change. *Harvard Business Review, 1*, 294–306.

Lantos, G. (2001). The boundaries of strategic corporate social responsibility. *Journal of Consumer Marketing, 18*(7), 595–630.

Lewin, K. (1947a). Frontiers in group dynamics. Concept, method and reality in social science; social equilibria and social change. *Human Relations, 1*(1), 5–41.

Lewin, K. (1947b). Group decision and social change. In T. Newcomb & E. Hartley (Eds.), *Readings in social psychology*. Holt, Reinhardt and Winstong, Inc.

Linnenluecke, M. K., & Griffiths, A. (2010). Corporate sustainability and organizational culture. *Journal of World Business, 45*(4), 357–366. https://doi.org/10.1016/j.jwb.2009.08.006

Linnenluecke, M. K., Russell, S. V., Griffi, A., Linnenluecke, M. K., & Griffiths, A. (2009). Subcultures and sustainability practices: The impact on understanding corporate sustainability. *Business Strategy and the Environment, 18*(December 2007), 432–452. https://doi.org/10.1002/bse

Lorenzi, N. M., & Riley, R. T. (2000). Managing change: An overview. *Journal of the American Medical Informatics Association, 7*(2), 116–124.

Lozano, R. (2006). Incorporation and institutionalization of SD into universities: Breaking through barriers to change. *Journal of Cleaner Production, 14*(9–11), 787–796. https://doi.org/10.1016/j.jclepro.2005.12.010

Lozano, R. (2012a). Orchestrating organisational changes for corporate sustainability: Overcoming barriers to change. *Greener Management International, 57*, 43–67.

Lozano, R. (2012b). Using game theory to develop stakeholder strategies towards more holistic corporate sustainability. In *Sustainable development*. New Research.

Lozano, R. (2012c). Towards better embedding sustainability into companies' systems: An analysis of voluntary corporate initiatives. *Journal of Cleaner Production, 25*(0), 14–26. https://doi.org/10.1016/j.jclepro.2011.11.060

Lozano, R. (2013). Are companies planning their Organisational changes for corporate sustainability? An analysis of three case studies on resistance to change and their strategies to overcome it. *Corporate Social Responsibility and Environmental Management, 20*(5), 275–295. https://doi.org/10.1002/csr.1290

Lozano, R. (2015). A holistic perspective on corporate sustainability drivers. *Corporate Social Responsibility and Environmental Management, 22*(1), 32–44. https://doi.org/10.1002/csr.1325

Lozano, R. (2018). Proposing a definition and a framework of Organisational sustainability: A review of efforts and a survey of approaches to change. *Sustainability, 10*(4), 1157. https://doi.org/10.3390/su10041157

Lozano, R. (2020). Analysing the use of tools, initiatives, and approaches to promote sustainability in corporations. *Corporate Social Responsibility and Environmental Management, 27*(2), 982–998. https://doi.org/10.1002/csr.1860

Lozano, R., & Barreiro-Gen, M. (2021). Disrupting the brave new world: COVID-19 effects on organisations' sustainability efforts. *Journal of Organizational Change Management.*, ahead-of-p(ahead-of-print). https://doi.org/10.1108/JOCM-09-2020-0276

Lozano, R., & Garcia, I. (2020). Scrutinizing sustainability change and its institutionalization in organizations. *Frontiers in Sustainability, 1*(May), 1–16. https://doi.org/10.3389/frsus.2020.00001

Lozano, R., & Huisingh, D. (2011). Inter-linking issues and dimensions in sustainability reporting. *Journal of Cleaner Production, 19*(2–3), 99–107. https://doi.org/10.1016/j.jclepro.2010.01.004

Lozano, R., Nummert, B., & Ceulemans, K. (2016). Elucidating the relationship between sustainability reporting and Organisational change Management for Sustainability. *Journal of Cleaner Production, 125*, 168–188. https://doi.org/10.1016/j.jclepro.2016.03.021

Lozano, R., & von Haartman, R. (2018). Reinforcing the holistic perspective of sustainability: Analysis of the importance of sustainability drivers in organizations. *Corporate Social Responsibility and Environmental Management, 25*(4), 508–522. https://doi.org/10.1002/csr.1475

Luthans, F. (2002). *Organizational behavior*. McGraw-Hill.

Machiavelli, N. (1966). The prince. The discourses.

MacLeod, S., & Lewis, D. (2004). Transnational corporations. Power, influence and responsibility. *Global Social Policy, 4*(I), 77–98.

Magretta, J. (2000). Growth through global sustainability. An interview with Monsanto's CEO Robert B. Shapiro. In *Harvard business review on business and the environment*. Harvard Business School Press.

Maurer, R. (1996). Beyond the wall of resistance. In *Unconventional strategies that build support for change*. Bard Book, Inc.

Oskarsson, K., & von Malmborg, F. (2005). Integrated management systems as a corporate response to sustainable development. *Corporate Social Responsibility and Environmental Management, 12*, 121–128.

Pettigrew, A., & Whipp, R. (1991). *Managing change for competitive success*. Blackwell Publishers Ltd.

Ragsdell, G. (2000). Engineering a paradigm shift? An holistic approach to organisational change management. *Journal of Organizational Change, 13*(2), 104–120.

Rogers, E. M. (1962). *Diffusion of innovations* (Fourth). The Free Press of Glencoe.

Rosner, W. J. (1995). Mental models for sustainability. *Journal of Cleaner Production, 3*(1–2), 107–121.

Salzmann, O., Ionescu-Somers, A. M., & Steger, U. (2005). The business case for corporate sustainability: Literature review and research options. *European Management Journal, 23*(1), 27–36. https://doi.org/10.1016/j.emj.2004.12.007

Sayce, S., Ellison, L., & Parnell, P. (2007). Understanding investment drivers for UK sustainable property. *Building Research and Information, 35*(6), 629–643. https://doi.org/10.1080/09613210701559515

Senge, P. M. (1999). The fifth discipline. In *The art & practice of the learning organization*. Random House Business Books.

Shephard, H. A. (1969). Innovation-resisting and innovation-producing organizations. In W. G. Bennis, K. D. Benne, & R. Chin (Eds.), *The planning of change* (Second). Holt, Rinehart and Winston, Inc.

Siebenhüner, B., & Arnold, M. (2007). Organizational learning to manage sustainable development. *Business Strategy and the Environment, 16*, 339–353.

van de Ven, A. H., Polley, D. E., Garud, R., & Venkataraman, S. (1999). *The innovation journey*. Oxford University Press.

Zadek, S. (1999). Stalking sustainability. *GMI, 26*, 1–11.

Chapter 6
Civil Society Organisations' Contributions to Sustainability

6.1 Introduction

Civil society organisations (CSOs) include educational institutions, religious organisations, non-governmental organisations (NGOs), and non-profit organisations. Of these, research on sustainability has taken place mostly by higher education institutions (HEIs), to a limited extent by religious organisations, and extremely limited or not at all by the remaining organisations (Lozano, 2018).

HEIs are the pinnacle of the academic system and are directly involved in the global knowledge network (Altbach et al., 2009). HEIs can serve as a proxy for other education institutions (Lozano, 2018). HEIs have, for centuries, been the places where young people have attended to received education and to gain knowledge through the university life experience (Elton, 2003; Lozano, 2010). HEIs educate the future professionals (Cortese, 2003), i.e. academics, decision makers, entrepreneurs, and leaders. HEIs can contribute actively to shape a sustainable future through the acquisition of knowledge, skills, attitudes, and values (UNESCO, 2017) through education and research. During the last 15 years, there have been a number of universities engaging with SD (Boks & Diehl, 2006; Wemmenhove & de Groot, 2001). Some of the ESD in HEIs efforts have included involvement in regional development (Dlouhá et al., 2013); reduction of greenhouse gas emissions (Klein-Banai & Theis, 2013); water conservation activities (Marinho et al., 2014); academic leadership commitment via the inclusion of SD in their mission and vision statements (Lee et al., 2013); signing of declarations, charters, and partnerships (Lozano, Lukman, et al., 2013; Lozano et al., 2014; Wright et al., 2004); environmental management systems (Holm et al., 2015; Nicolaides, 2006); Research on sustainability (Hugé et al., 2016); incorporating SD into curricula (Abdul-Wahab et al., 2003; Boks & Diehl, 2006; Davis et al., 2004; Thomas, 2004; von Blottnitz, 2006); and publication of sustainability reports (Ceulemans et al., 2015; Lozano, Llobet, & Tideswell, 2013).

R. Lozano, *Toward Sustainable Organisations*, Strategies for Sustainability,
https://doi.org/10.1007/978-3-030-99676-5_6

According to Beckford (2015), religious organisations focus on worship, prayer, meditation, teaching, healing, and spiritual well-being. They share much in common with other organisations and yet have characteristics that make them unique. They produce sacred knowledge, sacramental acts, transcendental experiences, prophecies, worship and meditation, healing rituals, rites of passage, religious ethics, and social welfare programmes. One of the first instances where the term "sustainable society" was used was at in 1974 at the World Council of Churches (Dresner, 2002). Accordingly, a sustainable society is defined by four principles: (1) social stability; (2) food supply, carrying capacity, and biodegradation; (3) reduction of non-renewable resources use; and (4) human activities that have little or no impact on the global climate. A recent development in religious organisations in contributing to sustainability was the Papal encyclical *Laudato Si'* (Francis, 2015). Nonetheless, there has been little academic research on these organisations' contributions to sustainability.

The following sections present the results of two surveys used to obtain primary data on how CSOs have been engaging and contributing to sustainability. The survey for the drivers was carried out between August 2016 and February 2017. The survey for the other categories was carried out between May and November 2018. The survey responses by each category are indicated by "*n*".

6.2 Sustainability Priorities and Impact

The CSO respondents came from higher education (26), NGOs (3), basic education (1), community based (1), faith based/religious (1), trusts/charity (1), and other (17).

The countries where the CSOs responses came from Spain (6), Germany (5), Netherlands (5), Austria (4), Belgium (3), Sweden (2), Switzerland (2), the UK (2), the USA (2), Argentina (1), Brazil (1), Canada (1), China (1), Finland (1), France (1), Gibraltar (1), India (1), Latvia (1), Luxembourg (1), Romania (1), Russia (1), Saudi Arabia (1), Serbia (1), Singapore (1), and the UAE (1).

Figure 6.1 shows the time that CSOs have been working with sustainability, where it can be seen that most of them have been working with it for more than 5 years (38% more than 15 years, 17% between 10 and 15 years, and 29% between 5 and 10 years).

Figure 6.2 shows the sustainability priorities of CSOs. As can be seen, the economic dimension is predominant, almost equally, in the short, medium, and long terms (76% very and extremely important in average). The foci on the environmental and social dimensions tend to be more important in the future, i.e. for the environmental dimension (71% for the long term and 58% for the medium term), about the same for the long and medium terms for the social dimension (78% and 81%, respectively) but lower in the short term (46% for the environmental dimension and 71% for the social one).

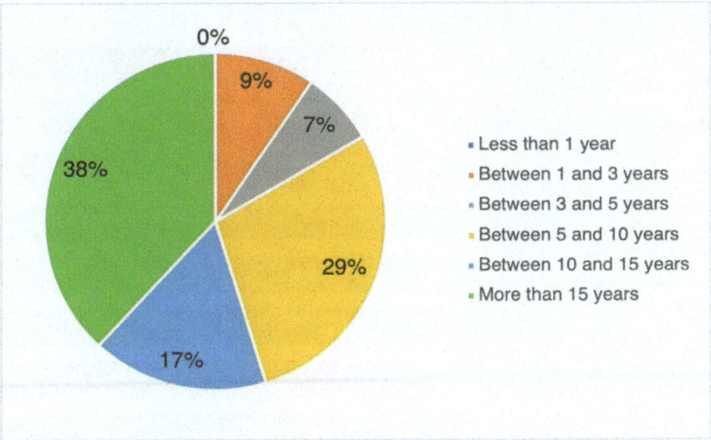

Fig. 6.1 Time working with sustainability ($n = 42$)

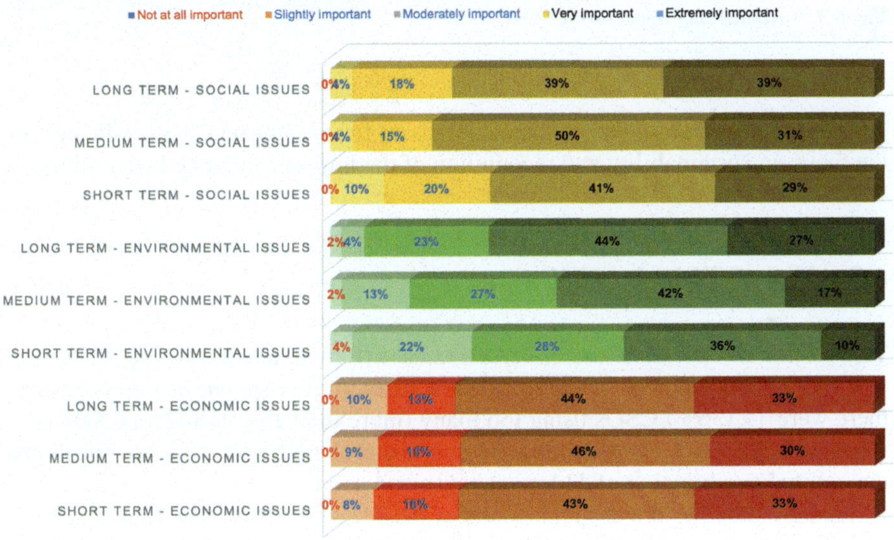

Fig. 6.2 Sustainability priorities of CSOs ($n = 52$)

Figure 6.3 shows the sustainability impacts of CSOs. As it can be seen, the economic dimension is either very or extremely important almost equally in the medium and long terms, with the short term having slightly less importance (between 59% and 71%). The environmental and social dimensions tend to be more important in the future, i.e. the long term (60% for the environmental dimension and 73% for the social one) is more important than the medium term (53% for the environmental dimension and 72% for the social one) compared to the short term (43% for the environmental dimension and 70% for the social one). These responses have a similar pattern to that of the focus ones (see Fig. 6.2), but the importance of impacts tends to be slightly less in the environmental and social dimensions, whereas for the

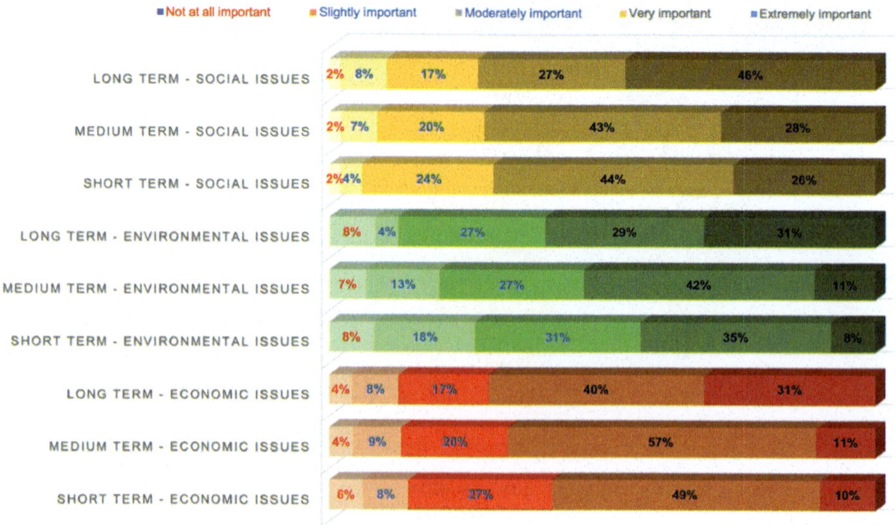

Fig. 6.3 Sustainability impacts of CSOs ($n = 52$)

economic dimension, the importance decreases for the impacts. This indicates that the focus on sustainability and recognition of the impacts could be better aligned.

6.3 Use of TIAs by CSOs

This section discusses the use of TIAs by CSOs (for explanations of the TIAs, refer to Chap. 3). An analysis of the number of TIAs used is shown in Fig. 6.4, where there were 19 (36%) CSOs using too many (more than 15), 21 (42%) CSOs using more than 7, 5 (10%) with too few (less than three), and 5 (10%) with, apparently, the optimal combination of TIAs, i.e. between 4 and 6.

Figure 6.5 shows the use and results of the TIAs by CSOs. The four TIAs most widely used have been CSR, corporate sustainability, cleaner production, and eco-efficiency. These are followed by corporate citizenship, design for the environment, and circular economy. The least known or used TIAS are TNS, sustainability reporting (SA8000), and sustainability reporting (AA1000).

Figure 6.6 presents the ratio of the results ("good" and "some", blue and green colours in Fig. 6.5) obtained by the TIAs versus the "no perceived results" (grey colour in Fig. 6.4) and "negative" results (red colour in Fig. 6.4). From these results, three groups appeared: high (with a ratio between 5 and 10), medium (with a ratio between 2 and 5), and low (with a ratio between 0 and 2). The high group has CSR and cleaner production. The medium group has corporate

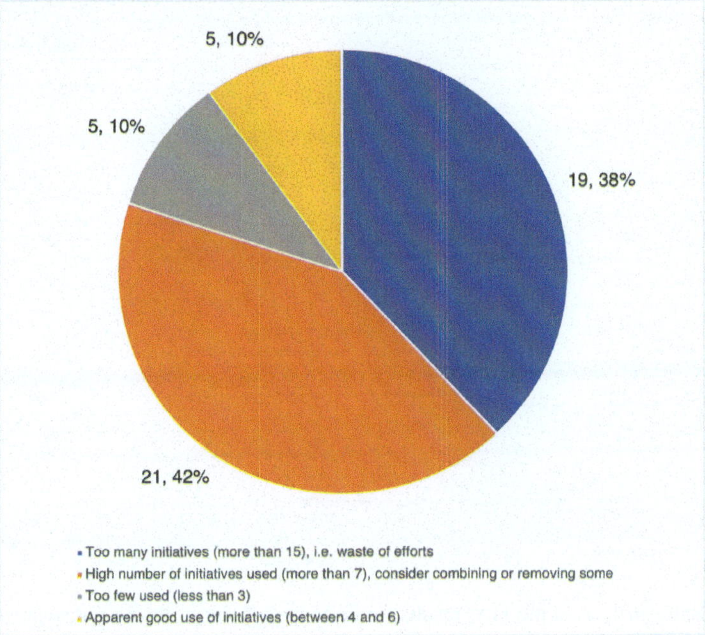

Fig. 6.4 Number of TIAs used by CSOs ($n = 86$)

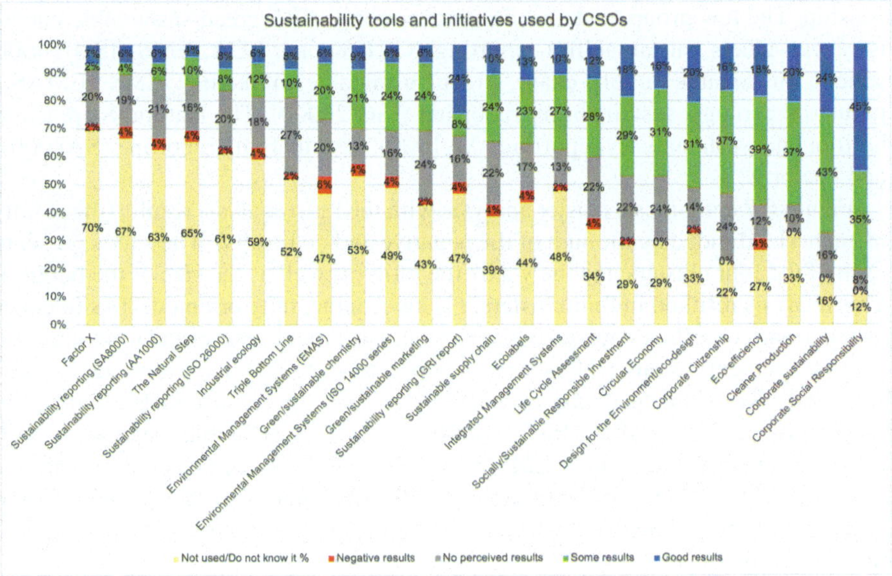

Fig. 6.5 Results obtained from the use of TIAs by CSOs ($n = 86$)

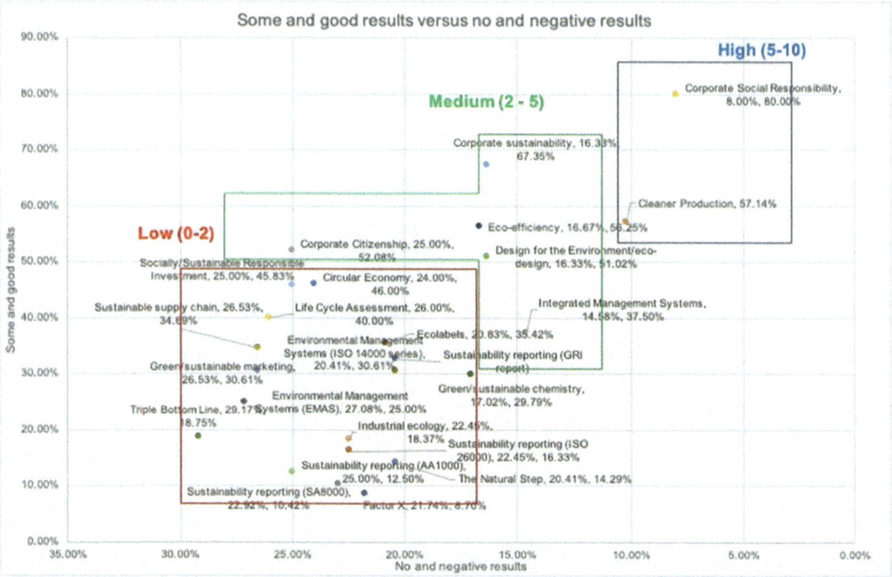

Fig. 6.6 Ratio analysis of the TIAs results (some and good) versus the no and negative results by CSOs ($n = 86$)

sustainability, eco-efficiency, design for the environment, IMS, and corporate citizenship. The low group contains circular economy, SRI, green/sustainable chemistry, eco-labels, sustainability reporting (GRI report), LCA, EMS (ISO 14000 series), sustainable supply chain, green/sustainable marketing, EMS (EMAS), industrial ecology, sustainability reporting (ISO 26000), The Natural Step, triple bottom line, sustainability reporting (AA1000), sustainability reporting (SA8000), and Factor X.

Figure 6.7 combines a cluster analysis with the ratio analysis results (the results of each TIA divided by the sum of the negative and no results). The x axis presents each of the four clusters and the y axis the ratio (as well as the size of the bubble). This analysis resulted in four clusters: (1) environmentally oriented TIAs (cleaner production, eco-efficiency, design for the environment, and circular economy,), (2) management-oriented TIAs (CSR, corporate sustainability, corporate citizenship, and SRI), (3) eclectically oriented TIAs (IMS, green/sustainable chemistry, eco-labels, LCA, EMS (ISO 14000 series), sustainable supply chain, green/sustainable marketing, EMS (EMAS), industrial ecology, sustainability reporting (ISO 26000), The Natural Step, TBL, sustainability reporting (AA1000), sustainability reporting (SA8000), and Factor X), and (4) reporting (sustainability reporting (GRI report)).

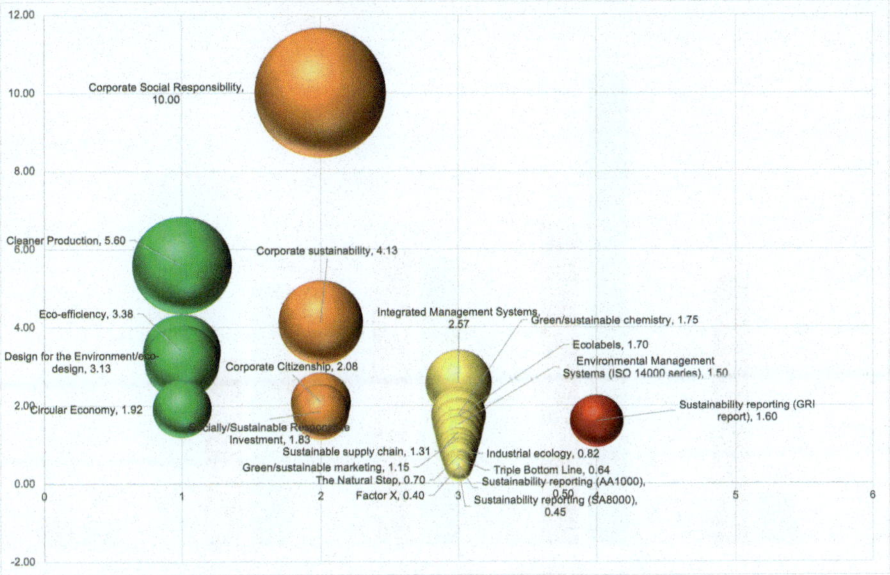

Fig. 6.7 Cluster and ratio analysis showing the results of the TIAs by CSOs in each cluster (depicted in the y axis and in the size of the bubble) ($n = 86$)

6.4 Sustainability Change Management in CSOs

This section discusses sustainability change management, as discussed in Chap. 4, in CSOs. The main driving forces for sustainability in CSOs have been (see Fig. 6.8) in descending ranking order: internal factors, mainly, but with some external stimuli; external stimuli and internal factors, equally; external stimuli, mainly, but with some internal factors; internal factors, solely; and external factors, solely.

The main resistance forces to sustainability in CSOs have been (see Fig. 6.8) in descending ranking order: internal factors, mainly, but with some external stimuli; external stimuli and internal factors equally; internal factors, mainly; external stimuli, mainly, but with some internal factors; and external stimuli, solely.

Figure 6.9 shows the ranking of the internal, connecting, and sustainability drivers in quintiles (the lowest on top and increasing clockwise in rank):

- Highest with two internal (proactive leadership and moral and ethical obligations), one connecting (reputation), and one external (increased levels of social awareness).
- High with one internal (organisation culture) and two external (national and regional policies and increased students' awareness).
- Middle with five internal (demands from employees, business case, sustainability champions, economic considerations, and risk avoiding), one connecting

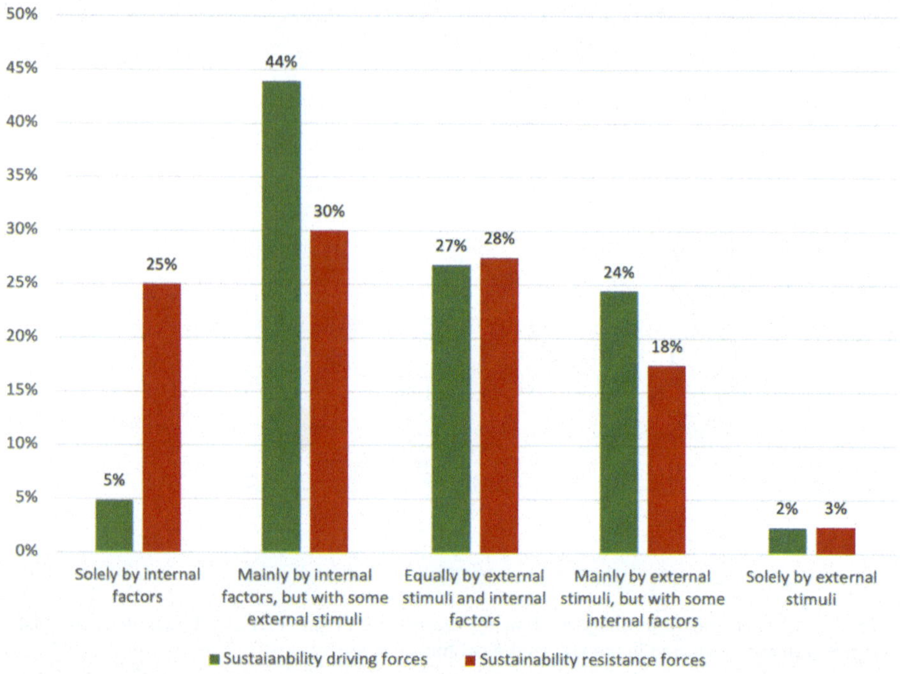

Fig. 6.8 CSOs sustainability driving and resistance forces ($n = 52$)

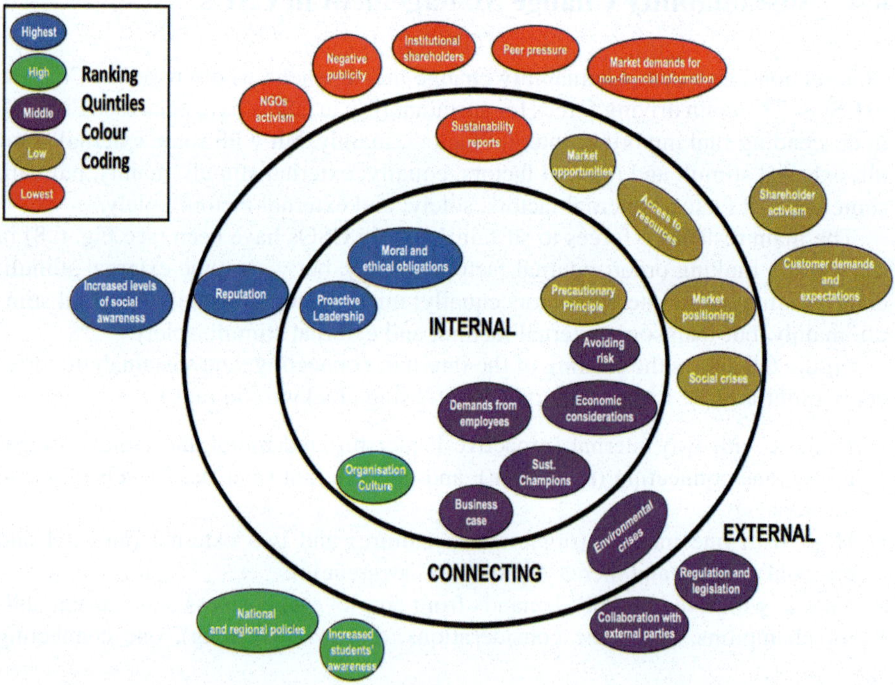

Fig. 6.9 Ranking of the internal, connecting, and external drivers in CSOs ($n = 108$). Source: adapted from (Lozano & von Haartman, 2018)

(environmental crises), and two external (collaboration with external parties and regulation and legislation).

- Lowers with one internal (precautionary principle), four connecting (social crises, market positioning, access to resources, and market opportunities), and two external (customer demands and expectations, and shareholder activism).
- Low with one connecting (sustainability reports) and five external (market demands for non-financial information, peer pressure, institutional shareholders, negative publicity, and NGOs activism).

Figure 6.10 shows the rankings of where sustainability changes have started in CSOs' systems elements (i.e. incorporation) and how they have been developed (i.e. institutionalised). The incorporation ranking shows that the changes have mainly started in service provision, organisational systems, and research and development. The institutionalisation ranking shows that the major changes have taken place in service provision, governance, and operations and production. The figure also shows the changes in ranking where two system elements had the same ranking in both stages (service provision and collaboration), three went up in ranking (governance, operations and production, and assessment and reporting), and four went down (research and development, supply chains, management and strategy, and organisational systems).

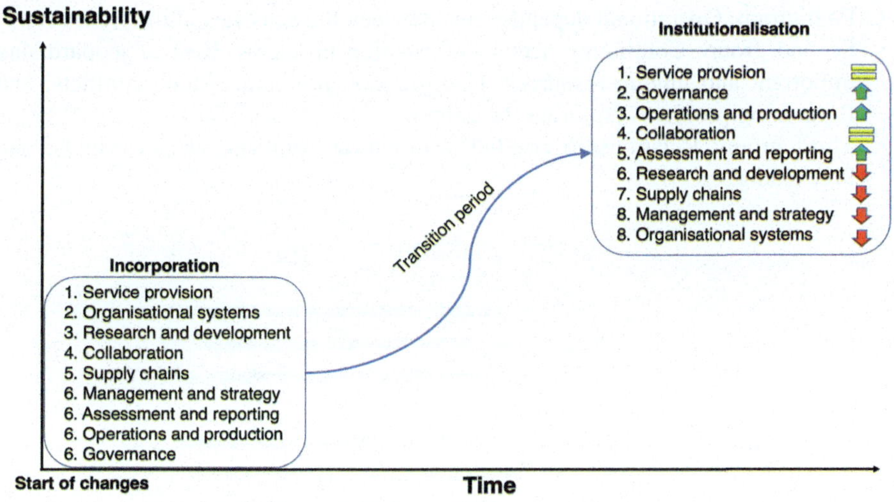

Fig. 6.10 Start of sustainability changes (incorporation) in the system elements and development of changes in the systems elements (institutionalisation) in CSOs ($n = 42$)

6.5 Sustainability Collaboration in CSOs

This section discusses sustainability collaboration, as discussed in Chap. 2, in CSOs. The respondents pointed out in which elements they collaborated and in what extent (Fig. 6.11). These can be divided into three groups:

- First group: Education, Innovation, and Services.
- Second group: Research, Recovery, Community, and Technology.
- Third group: Products and Business models.

The respondents also indicated the benefits accrued from collaboration. Those benefits were ranked in order of importance. They can be divided into three groups according to their ranking (see Fig. 6.12):

- First group: Enrich creativity and Transdisciplinary learning.
- Second group: More efficient processes and Access markets and knowledge.
- Third group: Ability to optimise financial and human capital, Offering benefits to all the players, reducing or removing conflicts, Being action orientated, and Decreased time needed to accomplish objectives.

The respondents also indicated the challenges that they had from sustainability collaboration. Those challenges were ranked in order of importance and divided into three groups (see Fig. 6.13):

- First group: Operational dependencies between the activities of different actors.
- Second group: Conflicting needs, Information problems, Risk of jeopardising important and unique resources, Coordination problems, Value conflicts, and Bargaining problems (splitting the gains).
- Third group: Relationship conflicts, Structural conflicts, Data conflicts, and Free riding.

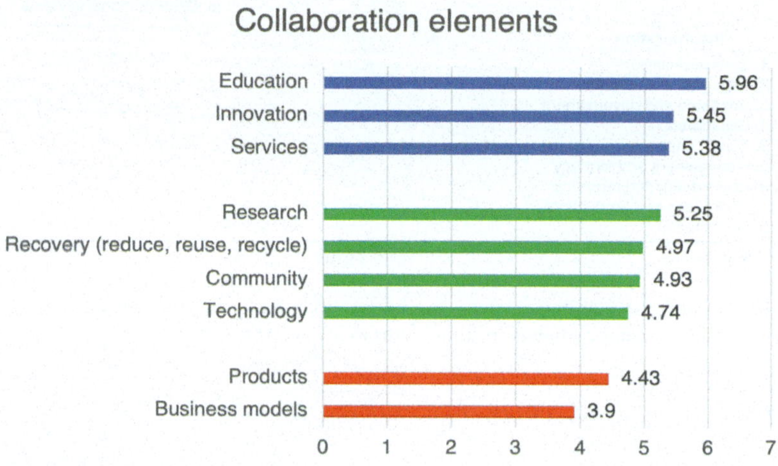

Fig. 6.11 Ranking of collaboration elements by CSOs ($n = 41$)

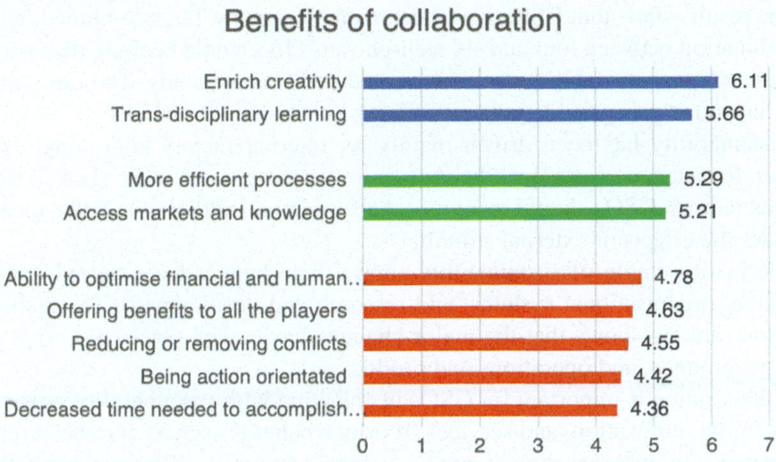

Fig. 6.12 Ranking of benefits obtained from collaboration by CSOs ($n = 41$)

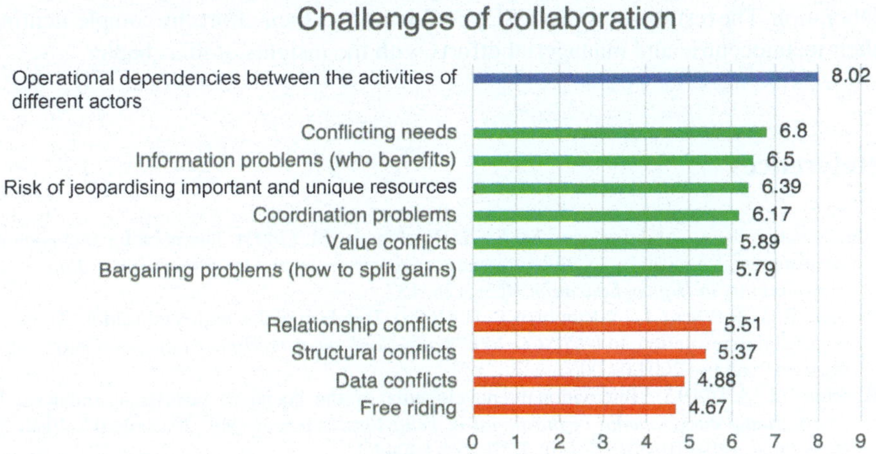

Fig. 6.13 Ranking of challenges that CSOs faced from sustainability collaboration ($n = 41$)

6.6 Discussion and Conclusions

CSOs have an important role in society by providing education, spiritual well-being, and other services not delivered by corporations and PSOs. CSOs have engaged with sustainability for a number of years, particularly HEIs, and some efforts by religious organisations. Research on other types of CSOs is still lacking. This chapter shows that CSOs' priorities and impacts on sustainability could be better aligned and particularly in aligning the short, medium, and long terms.

The results show that CSOs have been using too many TIAs combined, whereas a combination between four and six well-chosen TIAs would be more than adequate to address the sustainability dimensions and systems elements. The ratio analysis and cluster analysis provide guidance for such a choice.

Sustainability has been driven mainly by internal factors with some external stimuli. Resistance to sustainability has a similar pattern, but more skewed towards internal factors. CSOs should recognise the former and address the latter, as well as consider the effects of external stimuli.

The incorporation of sustainability shows that changes have started in service provision, organisational systems, and research and development. The institutionalisation ranking shows that the major changes have taken place in service provision, governance, and operations and production.

Collaboration is important for CSOs in engaging with sustainability, particularly in education, innovation, and services. It brings benefits such as enriched creativity and transdisciplinary learning; however, it entails a number of important challenges (e.g. operational dependencies).

This chapter provides insights on how CSOs have been engaging and contributing to sustainability with a particular focus on TIAs, change management, and collaboration. The results show that CSOs could improve their efforts by complementing their technocentric and managerial efforts with the insights of this chapter.

References

Abdul-Wahab, S. A., Abdulraheem, M. Y., & Hutchinson, M. (2003). The need for inclusion of environmental education in undergraduate engineering curricula. *International Journal of Sustainability in Higher Education, 4*(2), 126–137.

Altbach, P. G., Reisberg, L., & Rumbley, L. E. (2009). Trends in global higher education: Tracking an academic revolution. In *UNESCO 2009 World Conference on Higher Education*. https://doi.org/10.1016/j.bse.2004.04.006

Beckford, J. A. (2015). International encyclopedia of the social & behavioral sciences. In *International encyclopedia of the social & behavioral sciences* (Vol. 20, 2nd ed.). Elsevier. https://doi.org/10.1016/B978-0-08-097086-8.84044-1

Boks, C., & Diehl, J. C. (2006). Integration of sustainability in regular courses: Experiences in industrial design engineering. *Journal of Cleaner Production, 14*(9–11), 932–939.

Ceulemans, K., Lozano, R., & Alonso-Almeida, M. (2015). Sustainability reporting in higher education: Interconnecting the reporting process and organisational change Management for Sustainability. *Sustainability, 7*(7), 8881–8903. https://doi.org/10.3390/su7078881

Cortese, A. D. (2003). *The critical role of higher education in creating a sustainable future* (pp. 15–22). Planning for Higher Education.

Davis, S. A., Edmister, J. H., Sullivan, K., & West, C. K. (2004). Educating sustainable societies for the twenty-first century. *International Journal of Sustainability in Higher Education, 4*(2), 169–179. https://doi.org/10.1108/14676370310467177

Dlouhá, J., Huisingh, D., & Barton, A. (2013). Learning networks in higher education: Universities in search of making effective regional impacts. *Journal of Cleaner Production, 49*, 5–10. https://doi.org/10.1016/j.jclepro.2013.01.034

Dresner, S. (2002). *The principles of sustainability* (First). Earthscan Publications Ltd. www.earthscan.co.uk.

Elton, L. (2003). Dissemination of innovations in higher education: A change theory approach. *Tertiary Education and Management, 9*, 199–214.

Francis, P. (2015). Encyclical letter laudato si' of the holy father Francis on care for our common future. http://w2.vatican.va/content/francesco/en/encyclicals/documents/papa-francesco_20150524_enciclica-laudato-si.html.

Holm, T., Vuorisalo, T., & Sammalisto, K. (2015). Integrated management systems for enhancing education for sustainable development in universities: A memetic approach. *Journal of Cleaner Production, 106*, 155–163. https://doi.org/10.1016/j.jclepro.2014.03.048

Hugé, J., Block, T., Waas, T., Wright, T. S. A., & Dahdouh-Guebas, F. (2016). How to walk the talk? Developing actions for sustainability in academic research. *Journal of Cleaner Production, 137*, 83–92. https://doi.org/10.1016/j.jclepro.2016.07.010

Klein-Banai, C., & Theis, T. L. (2013). Quantitative analysis of factors affecting greenhouse gas emissions at institutions of higher education. *Journal of Cleaner Production, 48*, 29–38. https://doi.org/10.1016/j.jclepro.2011.06.004

Lee, K.-H., Barker, M., & Mouasher, A. (2013). Is it even espoused? An exploratory study of commitment to sustainability as evidenced in vision, mission, and graduate attribute statements in Australian universities. *Journal of Cleaner Production, 48*, 20–28. https://doi.org/10.1016/j.jclepro.2013.01.007

Lozano, R. (2010). Diffusion of sustainable development in universities' curricula: An empirical example from Cardiff University. *Journal of Cleaner Production, 18*(7), 637–644. https://doi.org/10.1016/j.jclepro.2009.07.005

Lozano, R. (2018). Proposing a definition and a framework of Organisational sustainability: A review of efforts and a survey of approaches to change. *Sustainability, 10*(4), 1157. https://doi.org/10.3390/su10041157

Lozano, R., Ceulemans, K., Alonso-Almeida, M., Huisingh, D., Lozano, F. J., Waas, T., Lambrechts, W., Lukman, R., & Hugé, J. (2014). A review of commitment and implementation of sustainable development in higher education: Results from a worldwide survey. *Journal of Cleaner Production, 108*. https://doi.org/10.1016/j.jclepro.2014.09.048

Lozano, R., Llobet, J., & Tideswell, G. (2013). The process of assessing and reporting sustainability at universities: Preparing the report of the University of Leeds. *Sostenibilidad, Tecnologia y Humanismo, 6*, 85–112.

Lozano, R., Lukman, R., Lozano, F. J., Huisingh, D., & Lambrechts, W. (2013). Declarations for sustainability in higher education: Becoming better leaders, through addressing the university system. *Journal of Cleaner Production, 48*, 10–19. https://doi.org/10.1016/j.jclepro.2011.10.006

Lozano, R., & von Haartman, R. (2018). Reinforcing the holistic perspective of sustainability: Analysis of the importance of sustainability drivers in organizations. *Corporate Social Responsibility and Environmental Management, 25*(4), 508–522. https://doi.org/10.1002/csr.1475

Marinho, M., Gonçalves, M. D. S., & Kiperstok, A. (2014). Water conservation as a tool to support sustainable practices in a Brazilian public university. *Journal of Cleaner Production, 62*, 98–106. https://doi.org/10.1016/j.jclepro.2013.06.053

Nicolaides, A. (2006). The implementation of environmental management towards sustainable universities and education for sustainable development as an ethical imperative. *International Journal of Sustainability in Higher Education, 7*(4), 414–424. https://doi.org/10.1108/14676370610702217

Thomas, I. (2004). Sustainability in tertiary curricula: What is stopping it happening? *International Journal of Sustainability in Higher Education, 5*(1), 33–47. https://doi.org/10.1108/14676370410517387

UNESCO. (2017). *Education for Sustainable Development*. http://en.unesco.org/themes/education-sustainable-development

von Blottnitz, H. (2006). Promoting active learning in sustainable development: Experiences from a 4th year chemical engineering course. *Journal of Cleaner Production, 14*(9–11), 916–923.

Wemmenhove, R., & de Groot, W. T. (2001). Principles for university curriculum greening. An empirical case study from Tanzania. *International Journal of Sustainability in Higher Education, 2*(3), 267–283.

Wright, Wright, T. S. A., & Wright. (2004). The evolution of environmental sustainability declarations in higher education. In A. E. J. Wals & P. B. Corcoran (Eds.), *Higher education and the challenge of sustainability: Problematics, promise, and practice* (pp. 7–19). Kluwer Academic Publishers.

Chapter 7
Corporations' Contributions to Sustainability

7.1 Introduction

Corporations,[1] in the modern sense, are institutions of associated individuals that, at least in Western countries, are legally, in power and liability, separated from their owners (Avi-Yonah, 2005; Diesendorf, 2000).

Traditionally, corporations have been defined as institutions dedicated to profitably producing and selling the goods and services that society demands (C.E.C., 2002; Demsetz, 1988; Freeman, 1984). Their main objective is to generate a satisfactory level of profit[2] for their legal owners (Argadoña, 1998; Boatright, 1996; Charreaux & Desbrières, 2001; Radin, 1932).

A growing number of companies have recognised the importance of contributing to sustainability (C.E.C., 2002; Elkington, 2002; WCED, 1987) and integrating sustainability into their strategic and operational decision-making processes (Thomas & Lamm, 2012). However, integrating sustainability principles into a company's system is challenging due to its complexity and the multidimensional issues (Langer & Schön, 2003).

The following sections present the results of two surveys used to obtain primary data on how corporations have been engaging and contributing to sustainability. The survey for the drivers was carried out between August 2016 and February 2017. The survey for the other categories was carried out between May and November 2018. The survey responses of each category are indicated as "*n*".

[1] Other terms used to refer to corporations include firms, enterprises, businesses, and companies.

[2] Profit is private benefit minus private cost (The Economist, 2005), where one of the highest costs for corporations is labour, which can range from 10 to 85 per cent (Farmer & Hogue, 1973). Although, labour might be the highest, without labour a corporation cannot exist or operate.

R. Lozano, *Toward Sustainable Organisations*, Strategies for Sustainability,
https://doi.org/10.1007/978-3-030-99676-5_7

7.2 Sustainability Priorities and Impact

The corporations respondents were from financial services (25), manufacturing (19), energy (17), food and beverages (12), chemicals (11), commercial services (10), construction (9), telecommunications (8), transport (8), computers and technology hardware (7), automotive (6), healthcare (5), forest and paper products (4), mining (4), and others (54).

The countries from which the corporation responses came were Sweden (27), Germany (23), Spain (18), the Netherlands (16), Italy (10), the UK (9), Finland (8), Portugal (8), Switzerland (8), Belgium (7), Turkey (7), Austria (6), Russia (5), the USA (5), Canada (4), France (4), Poland (4), Croatia (3), Czech Republic (3), the UAE (3), Denmark (2), Hungary (2), India (2), Norway (2), Serbia (2), South Africa (2), Botswana (1), Brazil (1), Chile (1), Greece (1), Ireland (1), Israel (1), Japan (1), Lebanon (1), Mexico (1), Sri Lanka (1), and Thailand (1).

Figure 7.1 shows the length of time that corporations have been working with sustainability. It can be seen that most of them have been engaged for more than 5 years (31% more than 15 years, 22% between 10 and 15 years, and 30% between 5 and 10 years).

Figure 7.2 shows the sustainability priorities of corporations. As can be seen, the economic dimension is predominant, almost equally, across the short, medium, and long terms (85% very or extremely important in average). The foci on the environmental and social dimensions tend to be more important for the long term (76% for the environmental dimension and 74% for the social) which is more important than the medium term (68% for the environmental dimension and 67% for the social) and least important for the short term (53% for the environmental dimension and 59% for the social).

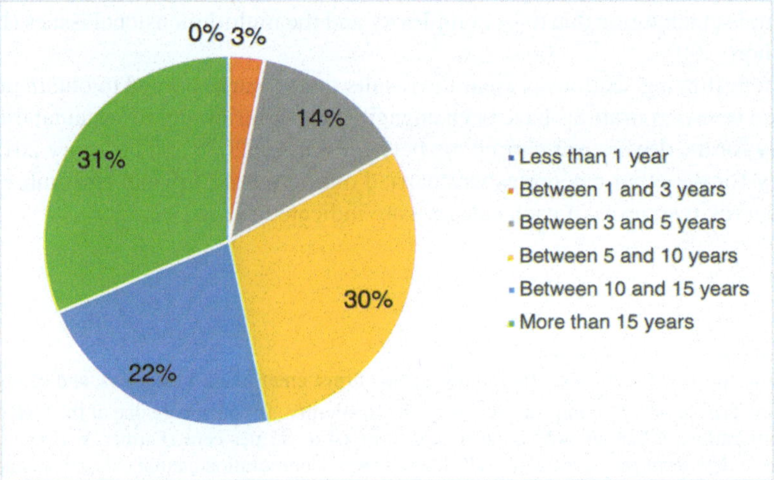

Fig. 7.1 Time over which corporations have been working with sustainability ($n = 195$)

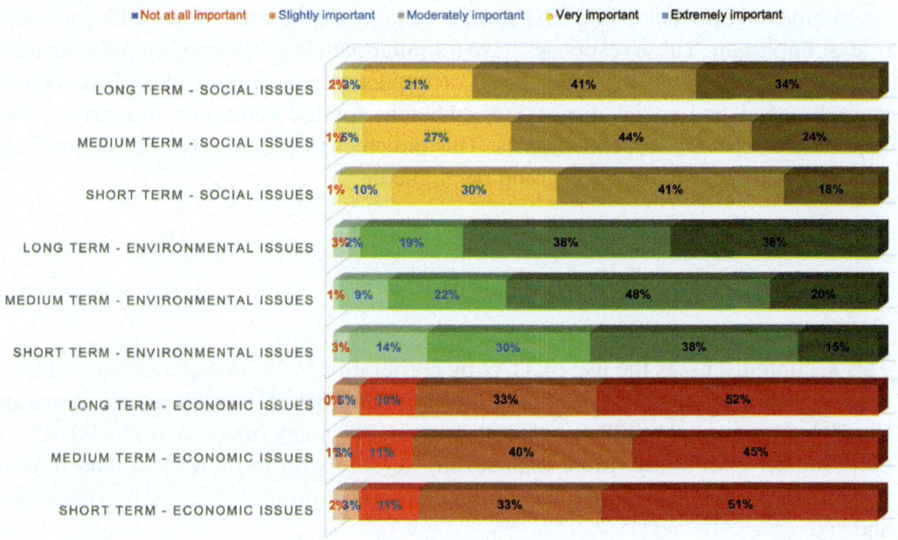

Fig. 7.2 Sustainability priorities of corporations ($n = 215$)

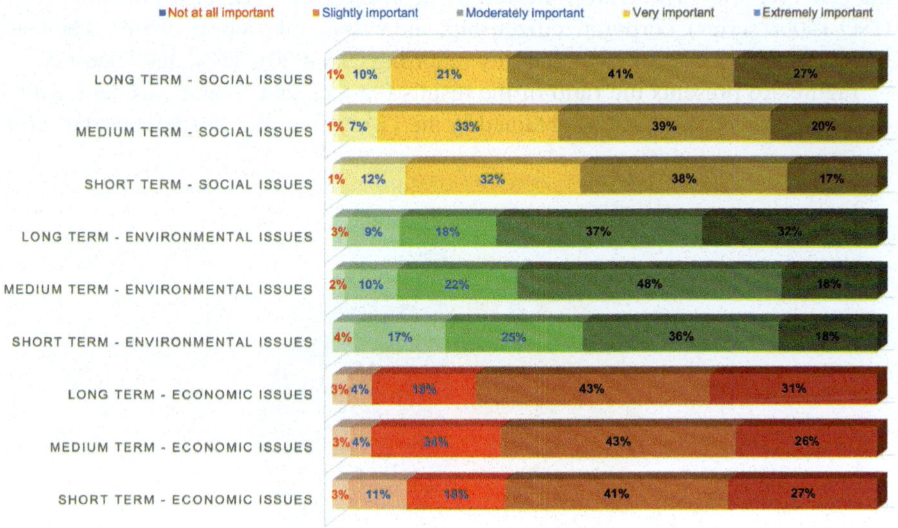

Fig. 7.3 Sustainability impacts of corporations (n = 215)

Figure 7.3 shows the sustainability impacts of corporations. As it can be seen, the economic dimension is either very or extremely important almost equally for the medium and long terms, with the short term having slightly less importance (between 68% and 74%). The environmental and social dimensions tend to be more important for the long term (70% for the environmental dimension and 68% for the social), and this is more important than the medium term (66% for the environmental dimension and 59% for the social), and the short-term (54% for both) is least important. These responses have a similar pattern to those of the focus examples (see Fig. 7.3), but the importance of impacts tends to be slightly less for the environmental and social dimensions, whereas for the economic dimension, the importance decreases for the impacts. This indicates that the focus on sustainability and recognition of the impacts could be aligned.

7.3 Use of TIAs by Corporations

This section discusses the use of TIAs by corporations (for an explanation of TIAs, refer to Chap. 3). An analysis was done on the number of TIAs used. As shown in Fig. 7.4, there were 94 (46%) corporations using too many (more than 15), 90 (45%) corporations with many (more than seven), 6 (3%) with too few (less than three), and 12 (6%) with, apparently, the optimal combination of TIAs, i.e. between four and six.

Figure 7.5 shows the use and results of the TIAs by corporations. The three TIAs most widely used have been CSR, corporate sustainability, and sustainability reporting (GRI reports). These are followed by eco-efficiency, cleaner production, EMS (ISO 14000 series), corporate citizenship, and sustainable supply chains. The least known or used TIAS are sustainability reporting (SA8000), TNS, and Factor X.

Figure 7.6 presents the ratio of the results ("good" and "some", as depicted by the blue and green in Fig. 7.5) obtained by the TIAs versus the "no perceived results"

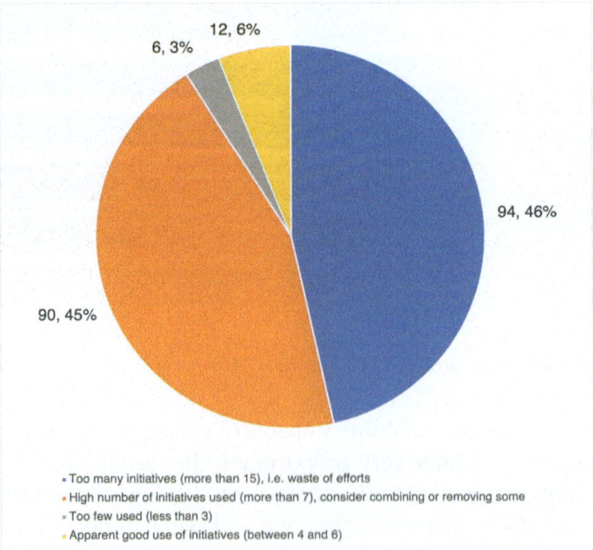

Fig. 7.4 Number of TIAs used by corporations ($n = 329$)

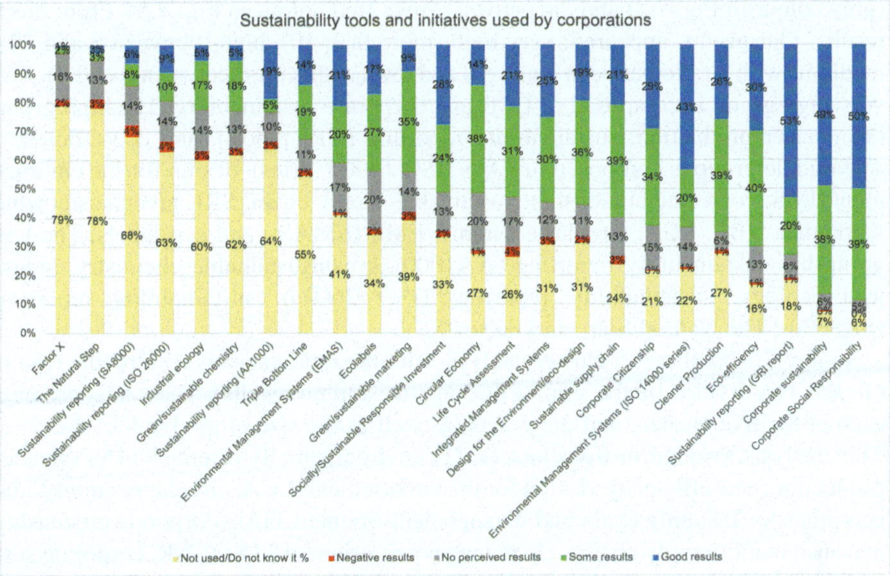

Fig. 7.5 Results obtained from the use of TIAs by corporations (n = 329). Source: adapted from (Lozano, 2020)

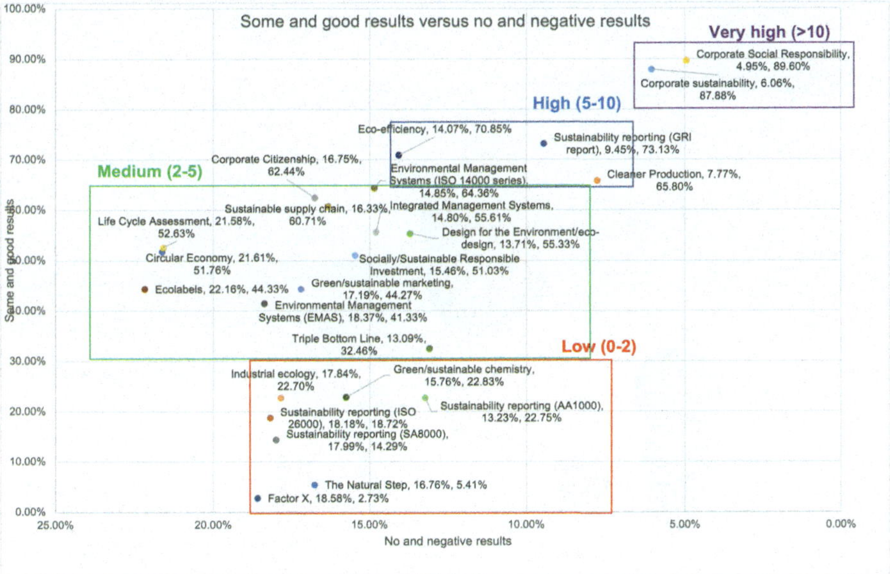

Fig. 7.6 Ratio analysis of the TIAs results (some and good) versus the no and negative results by corporations (n = 329). Source: adapted from (Lozano, 2020)

(grey colour in Fig. 7.5) and "negative" results (red colour in Fig. 7.5). From these results, four groups appeared: very high (more than 10), high (between 5 and 10), medium (with a ratio between 2 and 5), and low (with a ratio between 0 and 2). The very high group is composed of CSR and corporate sustainability. The high group has cleaner production, sustainability reporting (GRI report), and eco-efficiency. The middle group is composed of EMS (ISO 14000 series), design for the environment, IMS, corporate citizenship, sustainable supply chain, SRI, green/sustainable marketing, TBL, LCA, circular economy, EMS (EMAS), and eco-labels. The low group has sustainability reporting (AA1000), green/sustainable chemistry, industrial ecology, sustainability reporting (ISO 26000), sustainability reporting (SA8000), The Natural Step, and Factor X.

Figure 7.7 combines a cluster analysis with the ratio analysis results (the results of each TIA divided by the sum of the negative and no results). The x axis presents each of the five clusters and the y axis (as well as the size of the bubble) the ratio. This analysis resulted in five clusters: (1) environmentally oriented TIAs (cleaner production, eco-efficiency, design for the environment, LCA, circular economy, and eco-labels), (2) supply chain and management-oriented TIAs (corporate citizenship and sustainable supply chain), (3) management-oriented TIAs (CSR, corporate sustainability, and sustainability reporting (GRI report)), (4) environmental management-oriented TIAs (EMS (ISO 14000 series), IMS, and EMS (EMAS)), and (5) eclectically oriented TIAs (SRI, green/sustainable marketing, TBL, sustainability reporting (AA1000), green/sustainable chemistry, industrial ecology,

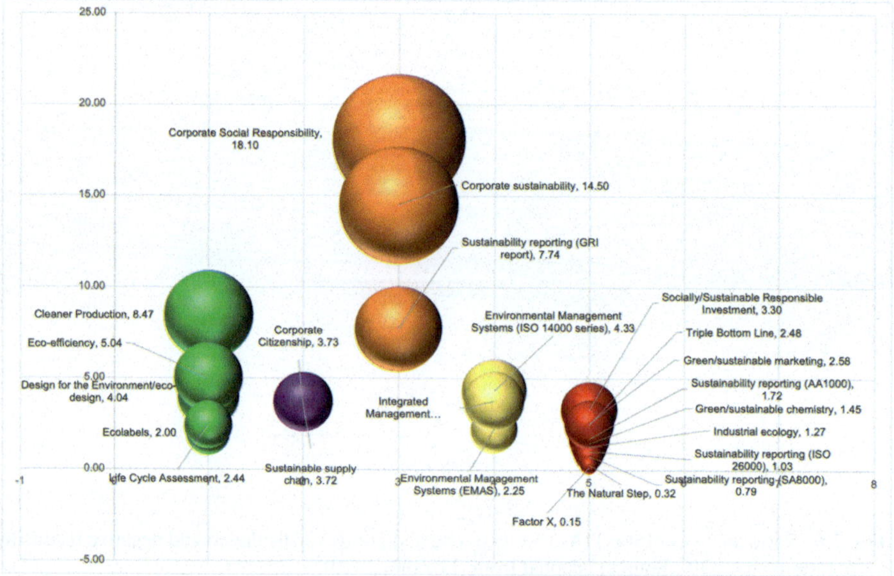

Fig. 7.7 Cluster and ratio analysis showing the results of the TIAs by corporations in each cluster (depicted in the y axis and in the size of the bubble) ($n = 329$). Source: (Lozano, 2020)

sustainability reporting (ISO 26000), sustainability reporting (SA8000), The Natural Step, and Factor X).

7.4 Sustainability Change Management in Corporations

This section discusses sustainability change management (refer to Chap. 4), in corporations. The main driving forces for sustainability in corporations have been (see Fig. 7.8) in descending ranking order: external stimuli and internal factors, equally, mainly by internal factors but with some external stimuli, mainly by external stimuli but with some internal factors, solely by internal factors, and solely by external factors.

The main resistance forces for sustainability in corporations have been (see Fig. 7.8), in descending ranking order: internal factors, mainly, but with some external stimuli; external stimuli and internal factors, equally; external stimuli, mainly, but with some internal factors; internal factors, solely; and external stimuli, solely.

Figure 7.9 shows the ranking of the internal, connecting, and sustainability drivers in quintiles (the lowest on top and increasing clockwise in rank):

- Highest with two internal (proactive leadership and organisation culture) and one connecting (reputation).
- High with three internal (moral and ethical obligations, avoiding risk, and precautionary principle), three connecting (market positioning, market opportunities,

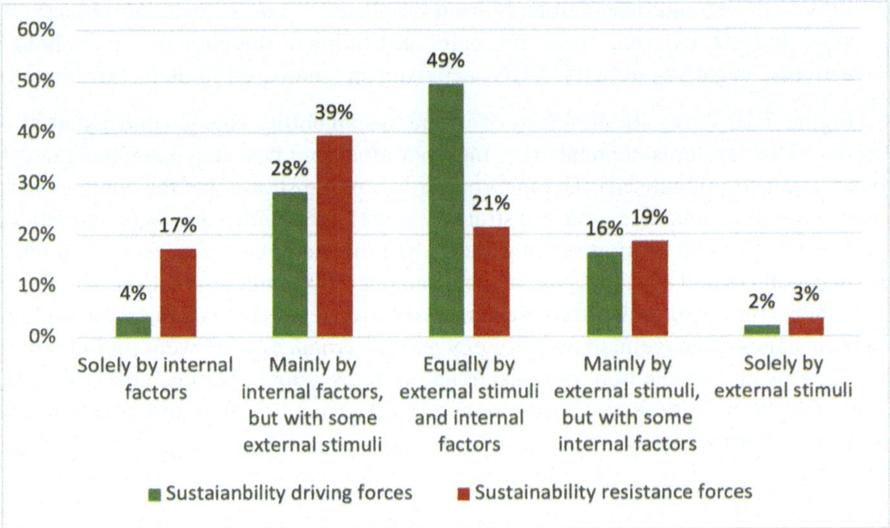

Fig. 7.8 Corporations' sustainability driving and resistance forces (n = 215)

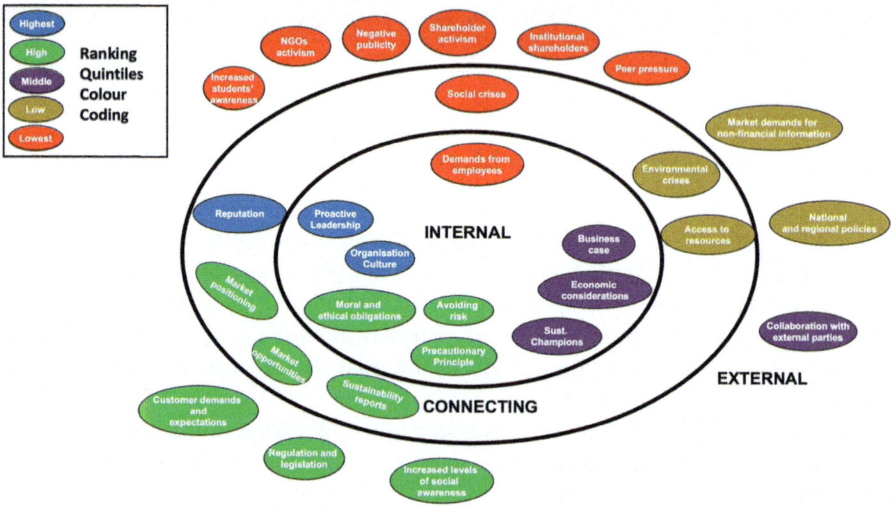

Fig. 7.9 Ranking of the internal, connecting, and external drivers in corporations (n = 108). Source: adapted from (Lozano & von Haartman, 2018)

and sustainability reports), and three external (customer demands and expectations, regulation and legislation, and increased levels of social awareness).

- Middle with three internal (sustainability champions, economic considerations, and business case) and one external (collaboration with external parties).
- Low with two connecting (access to resources environmental crises) and two external (national and regional policies and market demands for non-financial information).
- Low with one internal (demands from employees), one connecting (social crises), and six external (peer pressure, institutional shareholders, shareholder activism, negative publicity, NGOs activism, and increased students' awareness).

Figure 7.10 shows the rankings of where sustainability changes have started in corporations' systems elements (i.e. incorporation) and how they have been developed (i.e. institutionalised). The incorporation ranking shows that the changes have mainly started in management and strategy, governance, and assessment and reporting. The institutionalisation ranking shows that the major changes have taken place in management and strategy, governance, and operations and production. The figure also shows the changes in ranking where two system elements had the same ranking in both stages (management and strategy and governance), four went up in ranking (operations and production, organisational systems, supply chains, and collaboration), and three went down (assessment and reporting, research and development, and service provision).

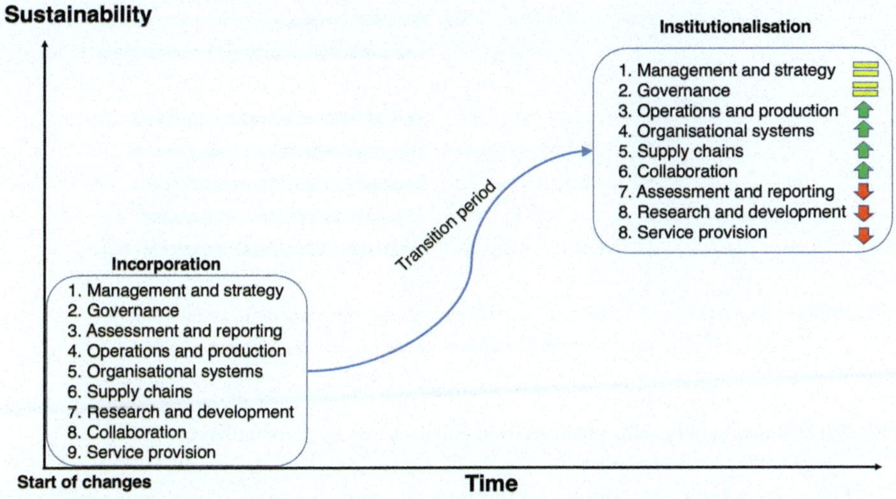

Fig. 7.10 Start of sustainability changes (incorporation) in the system elements and development of changes in the systems elements (institutionalisation) in corporations (n = 195)

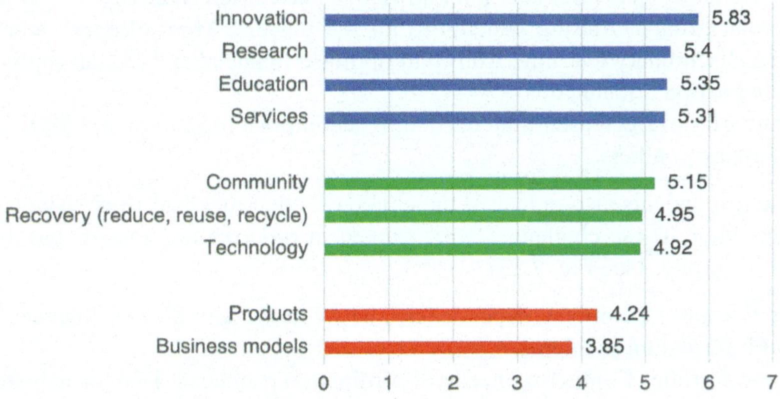

Fig. 7.11 Ranking of collaboration elements by corporations (n = 152)

7.5 Sustainability Collaboration in Corporations

This section discusses sustainability collaboration, as discussed in Chap. 2, in corporations. The respondents pointed out which elements they collaborated in and to what extent (Fig. 7.11). These can be divided into three groups:

- First group: Innovation, Research, Education, and Services.
- Second group: Community, Recovery, and Technology.
- Third group: Products and Business models.

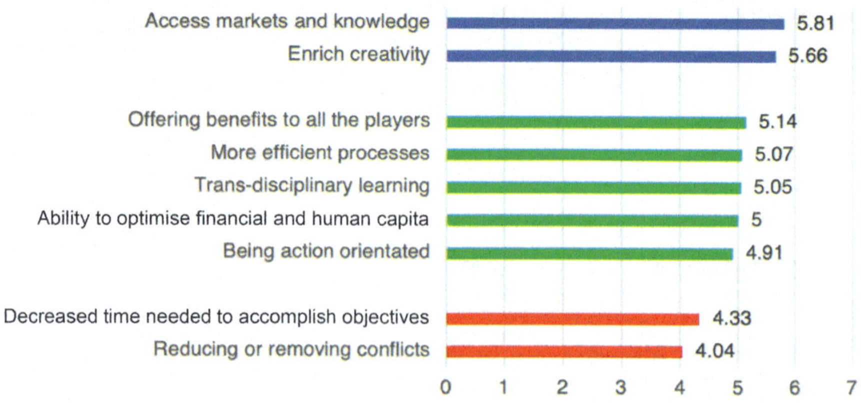

Fig. 7.12 Ranking of benefits obtained from collaboration by corporations ($n = 152$)

The respondents also indicated the benefits that they had. These were ranked in order of importance. They can be divided into three groups, according to their ranking (see Fig. 7.12):

- First group: Access markets and knowledge and Enrich creativity.
- Second group: Offering benefits to all the players, More efficient processes, Transdisciplinary learning, Ability to optimise financial and human capital, and Being action orientated.
- Third group: Decreased time needed to accomplish objectives and Reducing or removing conflicts.

The respondents also indicated the challenges that they had from sustainability collaboration. Those challenges were ranked in order of importance and divided into three groups (see Fig. 7.13):

- First group: Operational dependencies between the activities of different actors and Information problems.
- Second group: Conflicting needs, Coordination problems, Risk of jeopardising important and unique resources, and Value conflicts.
- Third group: Relationship conflicts, Data conflicts, Structural conflicts, Bargaining problems (splitting the gains), and Free riding.

7.6 Discussion and Conclusions

Corporations provide products and services in exchange for profit. Traditionally, their focus has been mainly on the economic dimension, but increasingly they are realising that they have an important role on the path to making societies more sustainable. Corporations have engaged with sustainability for a number of years, and research tends to be more prevalent with these organisations than CSOs and PSOs.

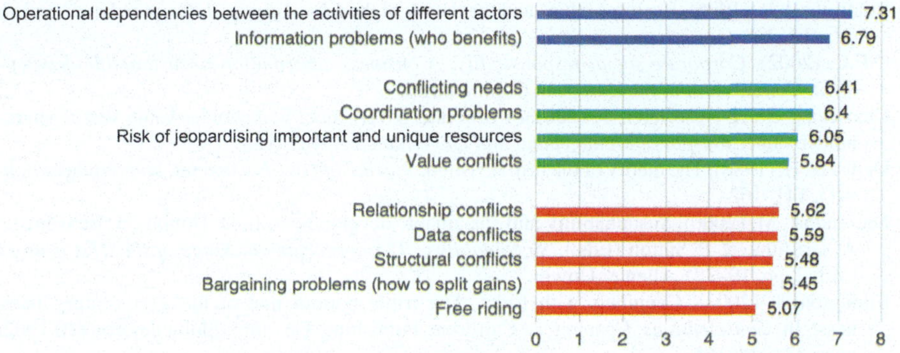

Fig. 7.13 Ranking of challenges that corporations faced from sustainability collaboration ($n = 152$)

This chapter shows that a corporation's priorities and impacts on sustainability could be better aligned, and particularly in aligning the short, medium, and long terms.

The results show that corporations have been using too many combined TIAs, whereas a combination of between four and six well-chosen TIAs would be more than adequate to address sustainability's dimensions and systems elements. The ratio analysis and cluster analysis provide guidance for such a choice.

Sustainability has been driven evenly by external stimuli and internal factors; however, the resistance to sustainability has been skewed towards internal factors. This shows a misalignment between the driving and resisting forces.

The incorporation of sustainability shows that changes have usually started in management and strategy, governance, and assessment and reporting. The last of these was overtaken by operations and production, organisational systems, supply chains, and collaboration during the institutionalisation.

Collaboration is important for corporations engaging with sustainability, particularly in the fields of innovation, research, education, and services. It brings benefits such as access to markets and enriched creativity; however, it entails a number of important challenges (e.g. operational dependencies, and information problems).

This chapter provides insights into how corporations have been engaging and contributing to sustainability, with a particular focus on TIAs, change management, and collaboration. The results show that corporations could improve their efforts by complementing their technocentric and managerial efforts using the insights of this chapter.

References

Argadoña, A. (1998). The stakeholder theory and the common good. *Journal of Business Ethics, 17*, 1093–1102.

Avi-Yonah, R. (2005). The cyclical transformations of the corporate form: A historical perspective on corporate social responsibility. *Delaware Journal of Corporate Law, 30*, 767–818. http://law.bepress.com/umichlwps-olin/art38/

Boatright, J. R. (1996). Business ethics and the theory of the firm. *American Business Law Journal, 34*, 217–238.
C.E.C. (2002). *Corporate social responsibility: A business contribution to sustainable development*. Commission of the European Communities.
Charreaux, G., & Desbrières, P. (2001). Corporate governance: Stakeholder value versus shareholder value. *Journal of Management and Governance, 5*, 107–128.
Demsetz, H. (1988). The theory of the firm revisited. *Journal of Law, Economics, and Organization, 4*(1), 141–162.
Diesendorf, M. (2000). Sustainability and sustainable development. In D. Dunphy, J. Benveniste, A. Griffiths, & P. Sutton (Eds.), *Sustainability: The corporate challenge of the 21st century* (Vol. 2, pp. 19–37). Allen & Unwin.
Elkington, J. (2002). Cannibals with forks. The triple bottom line of the 21st century business. In *Conscientious Commerce*. Capstone Publishing Co. http://doi.wiley.com/10.1002/tqem.3310080106
Farmer, R. N., & Hogue, W. D. (1973). *Corporate social responsibility*. Science Research Associate.
Freeman, R. E. (1984). *Strategic management: Stakeholder approach*. Pitman.
Langer, M. E., & Schön, A. (2003). Enhancing corporate sustainability. A framework based evaluation tools for sustainable development. Forschungsschwerpunkt Nachhaltigkeit und Umweltmanagement, Wirtschaftsuniversität Wien.
Lozano, R. (2020). Analysing the use of tools, initiatives, and approaches to promote sustainability in corporations. *Corporate Social Responsibility and Environmental Management, 27*(2), 982–998. https://doi.org/10.1002/csr.1860
Lozano, R., & von Haartman, R. (2018). Reinforcing the holistic perspective of sustainability: Analysis of the importance of sustainability drivers in organizations. *Corporate Social Responsibility and Environmental Management, 25*(4), 508–522. https://doi.org/10.1002/csr.1475
Radin, M. (1932). The endless problem of corporate personality. *Columbia Law Review, 32*, 643–667.
The Economist. (2005). *The good company. A survey of corporate social responsibility*. The Economist.
Thomas, T. E., & Lamm, E. (2012). Legitimacy and organizational sustainability. *Journal of Business Ethics, 110*(2), 191–203. https://doi.org/10.1007/s10551-012-1421-4
WCED. (1987). *Our common future (first)*. Oxford University Press.

Chapter 8
Public Sector Organisations' Contributions to Sustainability

8.1 Introduction

PSOs refer to any organisation that is under governmental control and that develops public goods or services (OECD, 2015). PSOs are major employers, providers of services, and consumers of resources (GRI, 2005) associated with significant aspects and impacts on the sustainability of the organisation. The public sector represents an important part of international economic activities (Ball & Grubnic, 2007; Walker & Brammer, 2012), for example, it accounts for about 20% of gross domestic product in the UK (Flynn, 2012). The government, whether central, regional, or local, performs a variety of roles, such as owner, funder, regulator, or purchaser of services provided by a PSO (Meyer & Leixnering, 2015). PSOs provide combinations of products and services that benefit society, such as schools, healthcare, or defending the country from military threats (Soyka, 2012).

The public sector influences all other sectors, given its size and related activities (Ball & Grubnic, 2007). It includes services that are provided as public goods, that are wholly or partly funded by taxation, and that are managed through governance arrangements that involve some degree of public accountability and political control or influence (Flynn, 2012).

Many PSOs have been focussing on becoming more sustainability-oriented (Soyka, 2012). Some of the sustainability efforts undertaken by PSOs include the adoption of social and environmental reports by Italian local authorities (Marcuccio & Steccolini, 2005), voluntary sustainability reporting based on the Global Reporting Initiative (GRI) Sustainability Reporting (SR) Guidelines (Guthrie & Farneti, 2008), sustainable public procurement (Andhov et al., 2020; Steurer & Berger, 2007; Witjes & Lozano, 2016), environmental disclosure practices in annual reports (Lynch, 2010), environmental reporting practices (Lodhia et al., 2012), performance measurement practice within government departments in Australia (Adams et al., 2014), and disclosure practices of sustainability information by

R. Lozano, *Toward Sustainable Organisations*, Strategies for Sustainability,
https://doi.org/10.1007/978-3-030-99676-5_8

European local governments (Navarro Galera et al., 2014). Despite these advances, PSOs are still lagging behind in SR (Dumay et al., 2010; Guthrie & Farneti, 2008; Lodhia et al., 2012), when compared to, for example, companies or higher education institutions (Lozano, 2018).

The following sections present the results of two surveys used to obtain primary data on how PSOs have been engaging and contributing to sustainability. The survey for the drivers was carried out between August 2016 and February 2017. The survey for the other categories was carried out between May and November 2018. The survey responses for each category are indicated as "*n*".

8.2 Sustainability Priorities and Impact

The PSO respondents were from education (16), environment (7), business (2), environment (7), external affairs (1), justice (1), municipal companies (1), transport (1), and others (13).

The countries where the PSOs responses came from were Germany (7), Spain (6), Sweden (5), Netherlands (5), Italy, (3) the USA (2), the UK (2), Finland (2), Belgium (2), Slovenia (1), Serbia (1), Qatar (1), Mexico (1), Bulgaria (1), Belarus (1), Austria (1), and Australia (1).

Figure 8.1 shows the length of time over which PSOs have been working with sustainability. It can be seen that most have been working with sustainability for

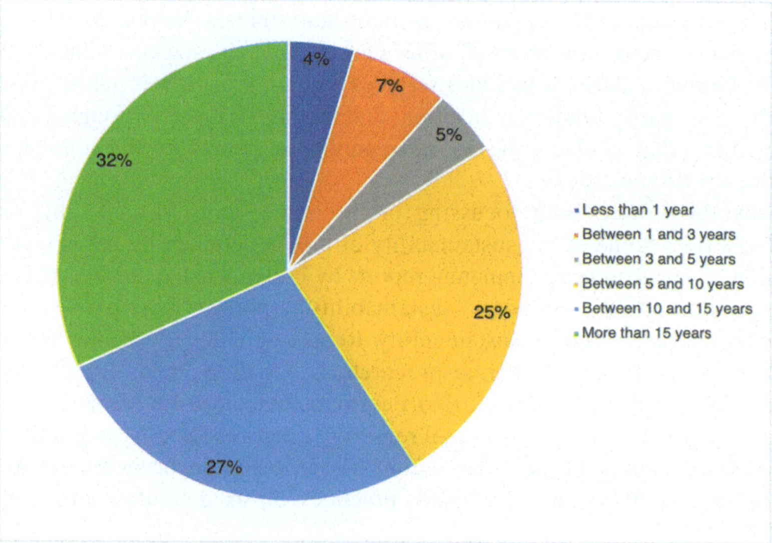

Fig. 8.1 Time that PSOs have been working with sustainability ($n = 44$)

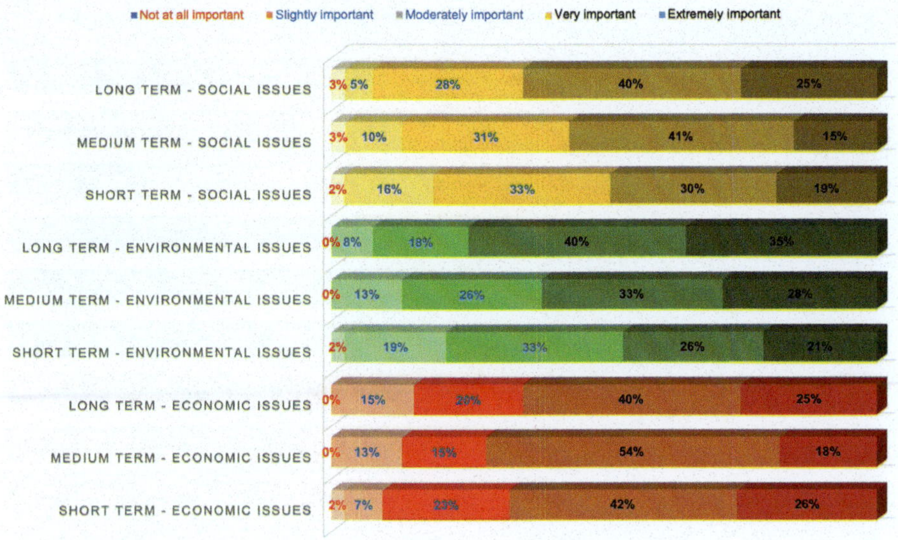

Fig. 8.2 Sustainability priorities of PSOs ($n = 45$)

more than 5 years (32% more than 15 years, 27% between 10 and 15 years, and 25% between 5 and 10 years).

Figure 8.2 shows the sustainability priorities of PSOs. As it can be seen, the economic dimension is predominant almost equally in the short, medium, and long terms (68% very and extremely important in average). The foci on the environmental and social dimensions tend to be more important in the future, i.e. the long term (75% for the environmental dimension and 65% for the social) is more important than the medium term (62% for the environmental dimension and 56% for the social) and least for the short term (47% for the environmental dimension and 49% for the social one).

Figure 8.3 shows the sustainability impacts of PSOs. As it can be seen, the economic dimension is either very or extremely important almost equal to the medium and long terms, with the short term having slightly less importance (between 48% and 51%). The environmental and social dimensions tend to be more important in the future, i.e. the long term (66% for the environmental dimension and 61% for the social) is more important than the medium term (59% for the environmental dimension and 51% for the social) and least for the short term (45% for both the environmental and social). These responses have a similar pattern to those for the focus responses (see Fig. 8.2), but the importance of impacts tends to be slightly less in the environmental and social dimensions, whereas, for the economic dimension, the importance decreases for the impacts. This indicates that the focus on sustainability and recognition of the impacts could be better aligned.

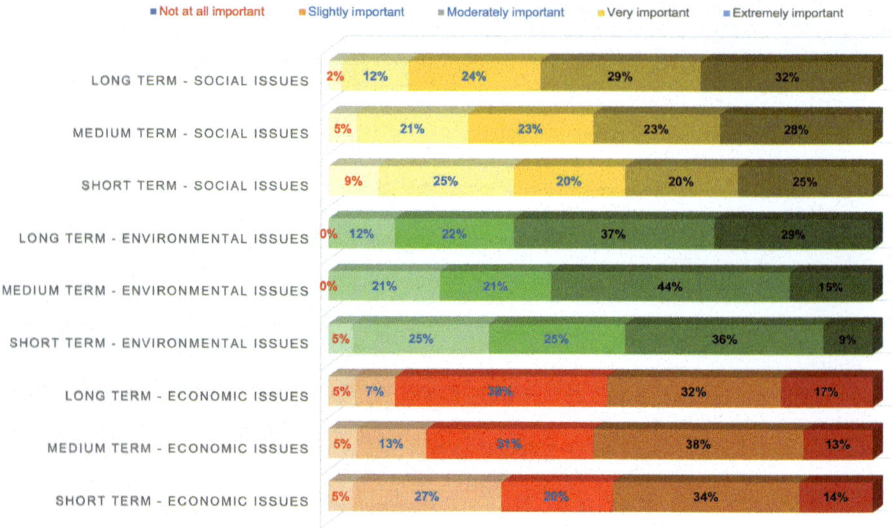

Fig. 8.3 Sustainability impacts of PSOs ($n = 45$)

8.3 Use of TIAs by PSOs

This section discusses the use of TIAs by PSOs (for an explanation of TIAs, refer to Chap. 3). An analysis was done on the number of TIAs used. As shown in Fig. 8.4, there were 18 (41%) PSOs using too many (more than 15), 17 (39%) with more than 7, 2 (4%) with too few (less than three), and 7 (16%) with, apparently, the optimal combination of TIAs, i.e. between 4 and 6.

Figure 8.5 shows the use of the TIAs by PSOs and the results. The three TIAs most widely used have been CSR, corporate sustainability, and eco-efficiency. These are followed by EMS (ISO 14000 series), cleaner production, LCA, EMS (EMAS), and design for the environment. The least known or used TIAS are TNS, sustainability reporting (SA8000), and Factor X.

Figure 8.6 presents the ratio of the results ("good" and "some", blue and green colours in Fig. 8.5) obtained by the TIAs versus the "no perceived results" (grey colour in Fig. 8.5) and "negative" results (red colour in Fig. 8.5). From these results, two groups appeared: medium (with a ratio between 2 and 5) and low (with a ratio between 0 and 2). The medium group contains EMS (EMAS), sustainability reporting (GRI report), CSR, corporate sustainability, cleaner production, EMS (ISO 14000 series), LCA, eco-efficiency, circular economy, and design for the environment. The low group contains corporate citizenship, TBL, green/sustainable marketing, eco-labels, IMS, sustainable supply chain, green/sustainable chemistry, industrial ecology, sustainability reporting (AA1000), sustainability reporting (ISO 26000), The Natural Step, sustainability reporting (SA8000), and Factor X.

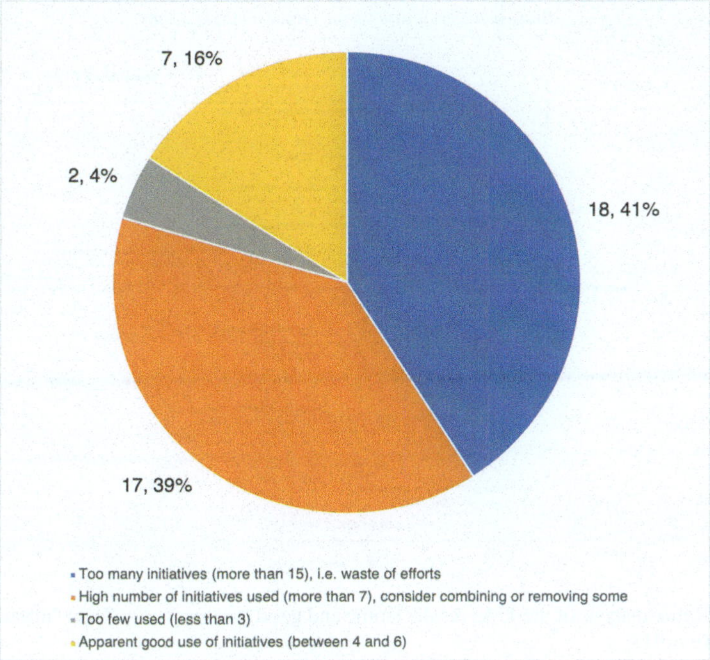

Fig. 8.4 Number of TIAs used by PSOs ($n = 91$)

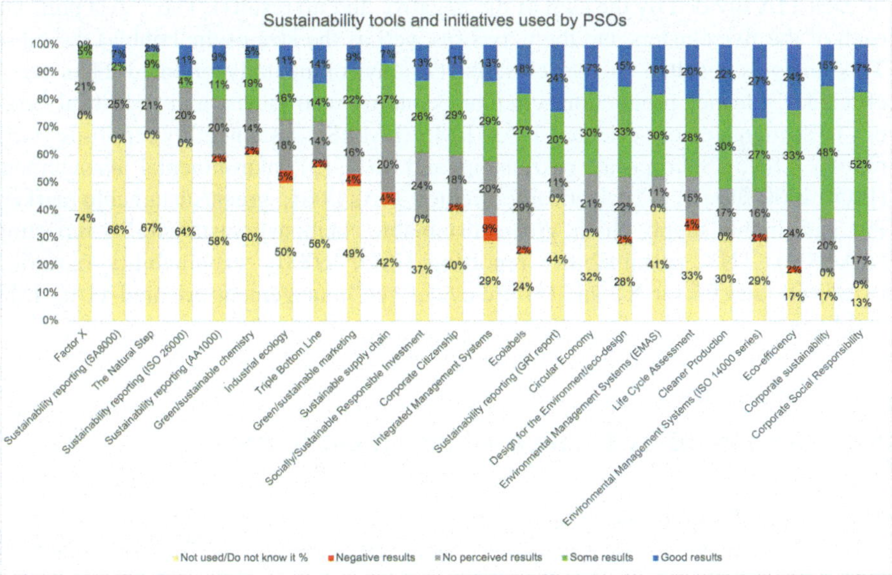

Fig. 8.5 Results obtained from the use of TIAs by PSOs ($n = 91$)

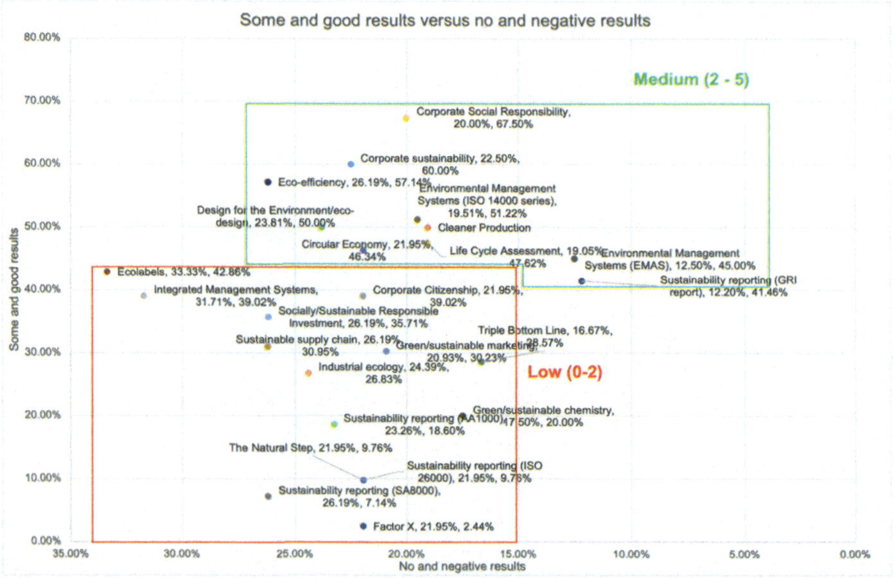

Fig. 8.6 Ratio analysis of the TIAs results (some and good) versus the "no" and "negative" results by PSOs ($n = 91$)

Figure 8.7 combines a cluster analysis with the ratio analysis results (the results of each TIA divided by the sum of the negative and no results). The x axis presents each of the five clusters and the y axis (as well as the size of the bubble) the ratio. This analysis resulted in five clusters: (1) environmentally oriented TIAs (eco-efficiency, cleaner production, circular economy, eco-labels, and industrial ecology), (2) management-oriented TIAs (EMS (EMAS), sustainability reporting (GRI report), CSR, CS, corporate citizenship, and SRI), (3) environmental management (ISO 14000 series), (4) eclectically oriented TIAs (TBL, green/sustainable marketing, sustainable supply chain, green/sustainable chemistry, sustainability reporting (AA1000), TNS, sustainability reporting (ISO 26000), sustainability reporting (SA8000), and Factor X), and (5) life cycle- and management-oriented TIAs (IMS and LCA).

8.4 Sustainability Change Management in PSOs

This section discusses sustainability change management, as discussed in Chap. 4, in PSOs. The main driving forces for sustainability in PSOs have been (see Fig. 8.8), in descending ranking order, due to external stimuli and internal factors, equally; external stimuli, mainly, but with some internal factors; and internal factors, mainly, but with some external stimuli.

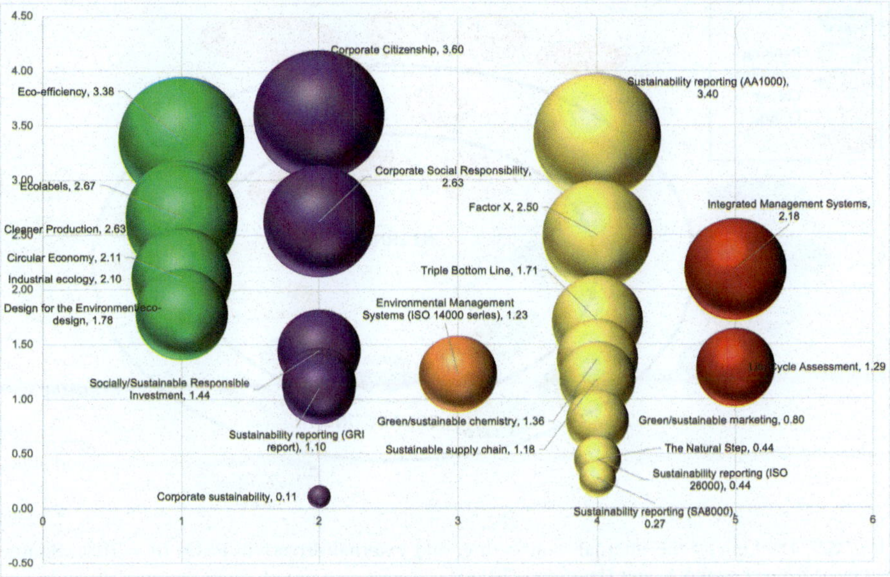

Fig. 8.7 Cluster and ratio analysis showing the results of the TIAs by PSOs in each cluster (depicted in the y axis and in the size of the bubble ($n = 91$)

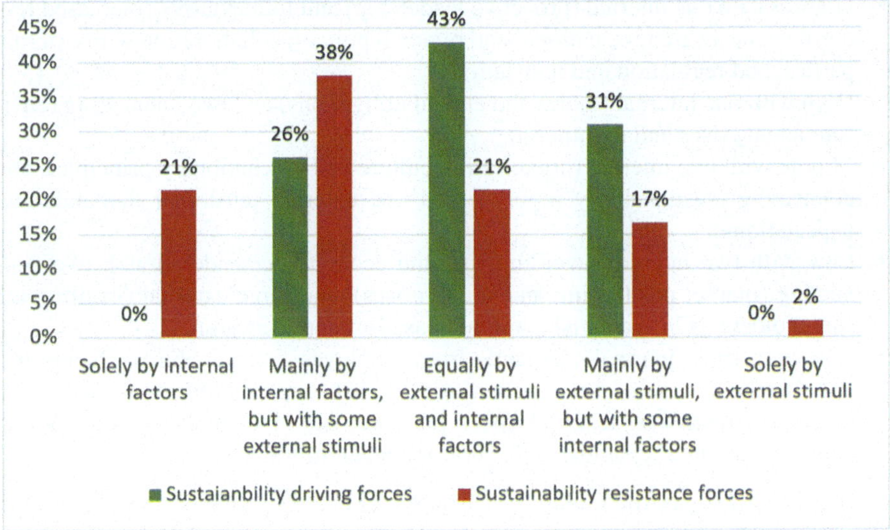

Fig. 8.8 PSOs sustainability driving and resistance forces ($n = 45$)

The main resistance forces for sustainability in PSOs have been (see Fig. 8.8), in descending ranking order, due to internal factors, mainly, but with some external stimuli; internal factors solely; external stimuli and internal factors, equally; external stimuli, mainly, but with some internal factors; and external stimuli, solely.

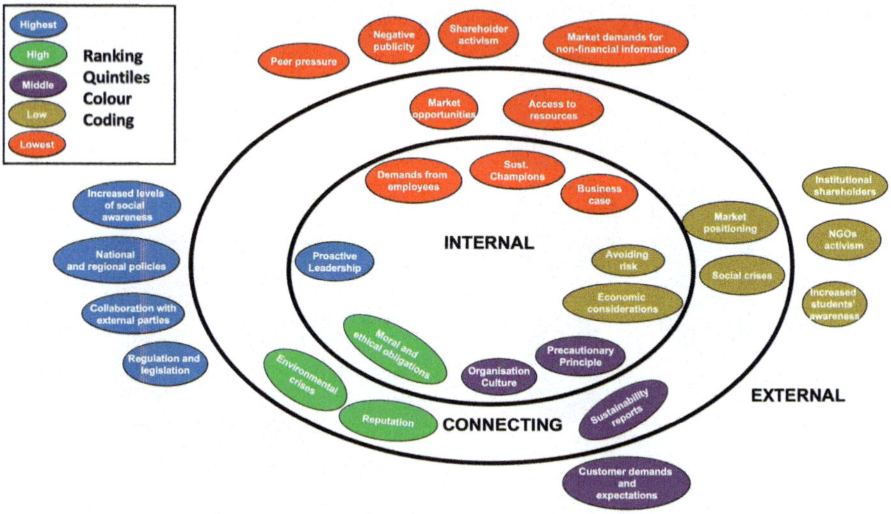

Fig. 8.9 Ranking of the internal, connecting, and external drivers in PSOs ($n = 108$). Source: adapted from (Lozano & von Haartman, 2018)

Figure 8.9 shows the ranking of the internal, connecting, and sustainability drivers in quintiles (the lowest on top and increasing clockwise in rank):

- Highest with one internal (proactive leadership) and four external (increased levels of social awareness, national and regional polices, collaboration with external parties, and regulation and legislation).
- High with one internal (moral and ethical obligations) and two connecting (environmental crises and reputation).
- Middle with two internal (organisation culture and precautionary principle), one connecting (sustainability reports), and one external (customer demands and expectations).
- Low with two internal (avoiding risk and economic considerations), two connecting (market positioning and social crises), and three external (institutional shareholders, NGOs activism, and increased students' awareness).
- Low with three internal (demands from employees, sustainability champions, and business case), two connecting (market opportunities and access to resources), and four external (peer pressure, negative publicity, shareholder activism, and market demands for non-financial information).

Figure 8.10 shows the rankings of where sustainability changes have started in PSOs' systems elements (i.e. incorporation) and how they have been developed (i.e. institutionalised). The incorporation ranking shows that the changes have started mainly in governance, management and strategy, and organisational systems. The institutionalisation ranking shows that the major changes have taken place in governance, management and strategy, and operations and production. The figure also shows the changes in ranking where three system elements had the same ranking in

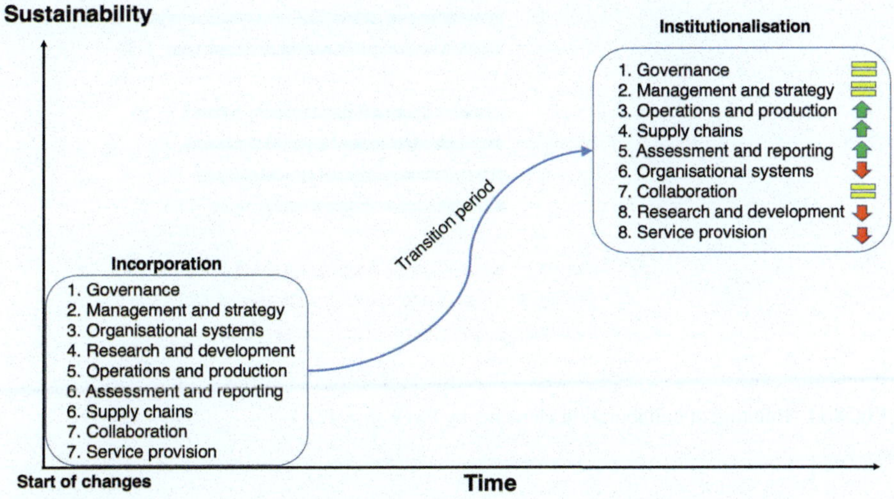

Fig. 8.10 Start of sustainability changes (incorporation) in the system elements and development of changes in the systems elements (institutionalisation) in PSOs ($n = 44$)

both stages (governance, management and strategy, and collaboration), three went up in ranking (operations and production, supply chains, and assessment and reporting), and three went down (organisational systems, research and development, and service provision).

8.5 Sustainability Collaboration in PSOs

This section discusses sustainability collaboration, as discussed in Chap. 2, in PSOs. The respondents pointed out in which elements they collaborated and in what extent (Fig. 8.11). These can be divided into three groups:

- First group: Innovation and Services.
- Second group: Education, Technology, Research, and Recovery.
- Third group: Community, Products, and Business models.

The respondents also indicated the benefits that they had. Those benefits were ranked in order of importance. They can be divided into three groups according to their ranking (see Fig. 8.12):

- First group: Transdisciplinary learning, Access markets and knowledge, and Offering benefits to all the players.
- Second group: Enrich creativity, More efficient processes, Being action-orientated.
- Third group: Ability to optimise financial and human capital, Decreased time needed to accomplish objectives, and Reducing or removing conflicts.

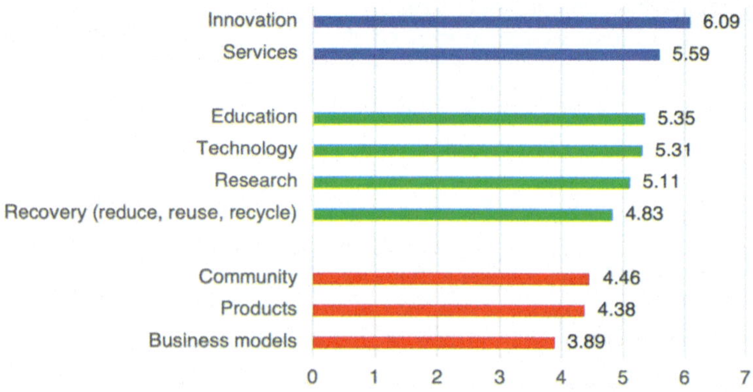

Fig. 8.11 Ranking of collaboration elements by PSOs ($n = 42$)

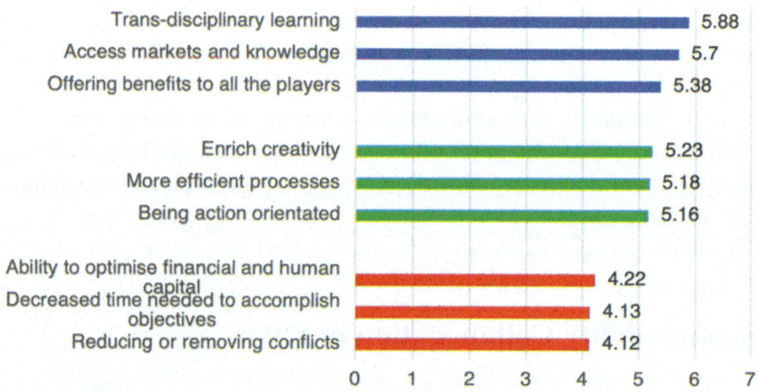

Fig. 8.12 Ranking of benefits obtained from collaboration by PSOs ($n = 42$)

The respondents also indicated the challenges that they had from sustainability collaboration. Those challenges were ranked in order of importance and divided into three groups (see Fig. 8.13):

- First group: Information problems, Conflicting needs, Operational dependencies between the activities of different actors, and Coordination problems.
- Second group: Data conflicts, Relationship conflicts, and Risk of jeopardising important and unique resources.
- Third group: Structural conflicts, Value conflicts, Bargaining problems (splitting the gains), and Free riding.

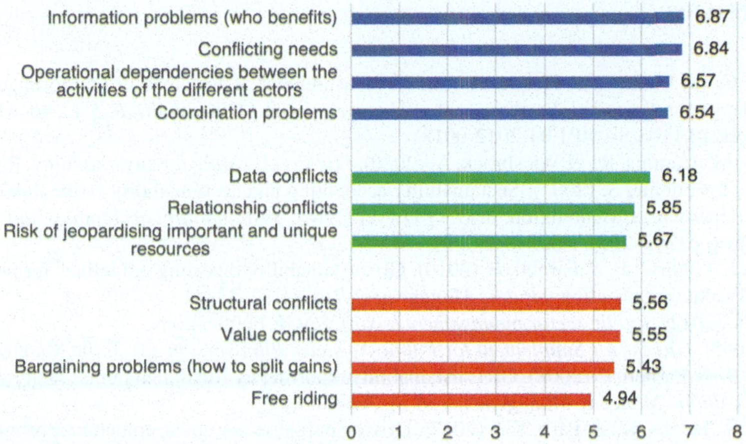

Fig. 8.13 Ranking of challenges that PSOs faced from sustainability collaboration ($n = 42$)

8.6 Discussion and Conclusions

PSOs play an important role in society by providing services and connecting the government to corporations and CSOs. Although PSOs have engaged with sustainability for a number of years, research on their efforts has been underrepresented when compared to corporations and CSOs. This chapter shows that PSOs' priorities and impacts on sustainability could be better aligned and particularly for aligning the short, medium, and long terms.

The results show that some PSOs have been using too many combined TIAs, whereas a blend of between four and six well-chosen TIAs would be more than adequate to address sustainability's dimensions and systems elements. The ratio analysis and cluster analysis provide guidance for making such a choice.

Sustainability has been driven evenly by external stimuli and internal factors; however, the resistance to sustainability has been skewed towards internal factors. This shows a misalignment between the driving and resisting forces.

The incorporation of sustainability shows that changes have started in governance, management and strategy, and organisational systems. The latter was overtaken by operations and production, supply chains, and assessment and reporting during the institutionalisation.

Collaboration is important for PSOs in engaging with sustainability, particularly in innovation and services. It brings benefits, such as transdisciplinary learning, access to markets, and offering benefits to all the players. However, it entails a number of important challenges (e.g. information problems, conflicting needs, operational dependencies, and coordination problems).

This chapter provides insights into how PSOs have been engaging and contributing to sustainability, with a particular focus on TIAs, change management, and collaboration. The results show that PSOs could improve their efforts by complementing their technocentric and managerial efforts with the insights of this chapter.

References

Adams, C. A., Muir, S., & Hoque, Z. (2014). Measurement of sustainability performance in the public sector. *Sustainability Accounting, Management and Policy Journal, 5*(1), 46–67. https://doi.org/10.1108/SAMPJ-04-2012-0018

Andhov, M., Caranta, R., & Wiesbrock, A. (2020). *Life-cycle costing for sustainability*. Routledge.

Ball, A., & Grubnic, S. (2007). Sustainability accounting and accountability in the public sector. In J. Unerman, J. Bebbington, & B. O'Dwyer (Eds.), *Sustainability accounting and accountability* (pp. 243–265). Routledge.

Dumay, J., Guthrie, J., & Farneti, F. (2010). Gri sustainability reporting guidelines for public and third sector organizations. *Public Management Review, 12*, 37–41.

Flynn, N. (2012). *Public sector management (sixth)*. SAGE Publications.

GRI. (2005). *GRI Sector Supplement for Public Agencies Pilot Version 1.0* (Issue March).

Guthrie, J., & Farneti, F. (2008). GRI sustainability reporting by Australian public sector organizations. *Public Money & Management, 28*(6), 361–366.

Lodhia, S., Jacobs, K., & Park, Y. J. (2012). Driving public sector environmental reporting. *Public Management Review, 14*(5), 631–647.

Lozano, R. (2018). Proposing a definition and a framework of Organisational sustainability: A review of efforts and a survey of approaches to change. *Sustainability, 10*(4), 1157. https://doi.org/10.3390/su10041157

Lozano, R., & von Haartman, R. (2018). Reinforcing the holistic perspective of sustainability: Analysis of the importance of sustainability drivers in organizations. *Corporate Social Responsibility and Environmental Management, 25*(4), 508–522. https://doi.org/10.1002/csr.1475

Lynch, B. (2010). An examination of environmental reporting by Australian state government departments. *Accounting Forum, 34*(1), 32–45. https://doi.org/10.1016/j.accfor.2009.11.001

Marcuccio, M., & Steccolini, I. (2005). Social and environmental reporting in local authorities. *Public Management Review, 7*(2), 155–176.

Meyer, R. E., & Leixnering, S. (2015). Public sector organizations. In *International Encyclopedia of the Social & Behavioral Sciences* (Vol. 18, 2nd ed.). Elsevier. https://doi.org/10.1016/B978-0-08-097086-8.73056-X

Navarro Galera, A., de los Ríos Berjillos, A., Ruiz Lozano, M., & Tirado Valencia, P. (2014). Transparency of sustainability information in local governments: English-speaking and Nordic cross-country analysis. *Journal of Cleaner Production, 64*, 495–504. https://doi.org/10.1016/j.jclepro.2013.07.038

OECD. (2015). *Government at a glance 2015*. OECD Publishing.

Soyka, P. A. (2012). *Creating a sustainable organization*. FT Press.

Steurer, R., & Berger, G. (2007). Sustainable public procurement in EU member states: Overview of government initiatives and selected cases. Final report to the EU high-level group on CSR. *European Commission, ..., October*. http://www.sustainability.at/pdf/csr/policies/Sustainable Public Procurement in EU Member States_Final Report.pdf.

Walker, H., & Brammer, S. (2012). The relationship between sustainable procurement and e-procurement in the public sector. *International Journal of Production Economics, 140*(1), 256–268. https://doi.org/10.1016/j.ijpe.2012.01.008

Witjes, S., & Lozano, R. (2016). Sustainable public procurement: Moving to more sustainable societies. *Journal of Cleaner Production*, 85–95. Virtual sp.

Chapter 9
Towards More Sustainable Organisations: Discussions and Conclusions

9.1 Introduction

Organisations are an integral part of modern societies (Scott & Davis, 2015). They are subsystems of a larger environmental system (Porter et al., 1975; Stacey, 1993) and have boundaries between the inside and the outside of the organisation (Scott & Davis, 2015). Organisations are affected by internal and external forces and conditions (Gebauer & Worch, 2015; Jones, 2013).

Organisations are, usually, divided into (Holliday et al., 2002) civil society organisations (CSOs), (including society at large, education institutions, and non-governmental organisations), corporations (small, medium, large, national, international, and transnational), and public sector organisations (PSOs) (local, regional, and national).

Organisations are complex social systems with sets of interrelated units engaged in joint problem-solving to accomplish a goal or objective (Jones, 2013; Porter et al., 1975; Rogers, 1962) and to create value and acceptable outcomes for stakeholders (Jones, 2013).

Organisations have been instrumental in driving sustainability (Danter et al., 2000; Holliday et al., 2002; Jennings & Zandbergen, 1995), where the importance of sustainability's dimensions depends on an organisation's nature and purpose (Soyka, 2012). The impacts of organisations have been the highest on the social dimension (67%), followed by the economic (61%), and then the environmental (55%) (Lozano, 2018).

In the last decade, there has been an increasing interest in organisational sustainability (Corsi & Arru, 2020; Hörisch et al., 2015; Lozano, 2008, 2018; Pfeffer, 2010). This book is based on the following definition of "organisational sustainability" (illustrated in Fig. 9.1): "The contributions of the organisation to sustainability equilibria, including the economic, environmental, and social dimensions of today, as well as their interrelations within and throughout the time dimension (i.e., the

R. Lozano, *Toward Sustainable Organisations*, Strategies for Sustainability,
https://doi.org/10.1007/978-3-030-99676-5_9

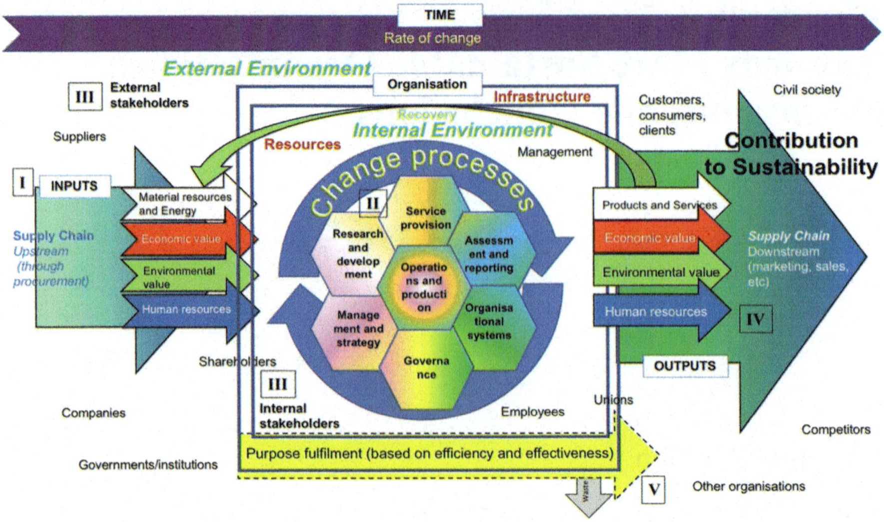

Fig. 9.1 Organisational sustainability framework. Source: (Lozano, 2018)

short-, long-, and longer-term). This entails the continuous incorporation and implementation of sustainability issues in the organisation's system elements (operations and production, management and strategy, governance, organisational systems, service provision, and assessment and reporting), as well as change processes and their rate of change. The system elements and change processes transform the inputs (in respect of material and resources that have economic, environmental, and social value) into outputs (products, services, and waste, with their economic, environmental, and social value). These fulfil the organisation's goal or objective, based on resource efficiency and effectiveness. The organisation is affected by the organisation's non-human and human resources (i.e., individuals, groups, culture, values, attitudes, and norms), its infrastructure, its supply chain (upstream and downstream), and the interactions with its stakeholders (internal, interconnecting, and external)" (Lozano, 2018).

Most of the literature on organisational sustainability has focused on corporations (Dyllick & Hockerts, 2002; Pfeffer, 2010; Schaltegger & Burrit, 2005; Tschopp, 2005), followed by education institutions, in particular higher education (Dlouhá et al., 2013; Lozano et al., 2014). Limited, yet increasing, attention has been directed to PSOs and other civil society organisations (Domingues et al., 2017; Dumay et al., 2010; Guthrie & Farneti, 2008; Lodhia et al., 2012). Most organisational sustainability efforts have focused on "technocentric" (Doppelt, 2003; Linnenluecke et al., 2009) or managerial ploys (Henriques & Richardson, 2005), and there has been increasing interest in organisational systems and change processes (Lozano, 2012). However, incorporating, integrating, and implementing sustainability in organisations is a complex and multidimensional process (Langer & Schön, 2003; Lozano, 2018).

9.2 Sustainability Priorities and Impacts

This section compares the sustainability priorities and impacts, as presented in Chapters 1, 5, 6, and 7, between corporations, CSOs, and PSOs. The organisations analysed in this book have been working with sustainability for more than 5 years (between 31% and 38% more than 15 years, between 17% and 27% between 10 and 15 years, and between 25% and 30% between 5 and 10 years).

The three types of organisations mentioned above show a balanced pattern between their sustainability priorities in the short term (as shown in Fig. 9.2). Corporations consider the short term to be considerably important (around 85% for each dimension); for CSOs the importance in the medium term is around 75%; and for PSOs it is between 65% and 72%. The environmental dimension tends to be less important for the three types (46% for CSOs, 47% for PSOs, and 53% for corporations) in the short term but increases with a longer-term perspective (between 58% for CSOs and 68% for corporations) and over 70% in the long term. The social dimension shows a similar pattern in the three types, where it is considered more important in PSOs (between 65% in the long term to 74% in the short term and 78% in the medium term) than for CSOs (with 56% in the long term, 67% in the short term, and 81% in the medium term), and corporations (with 49% in the long term, 59% in the short term, and 71% in the medium term).

In contrast, the impacts on sustainability presented different changes with respect to organisation type (see Fig. 9.3). PSOs and CSOs have a similar pattern in the economic dimension, both with greater impacts in the short and medium terms than in the long term, but PSOs give slightly higher numbers (74% in the short term vs

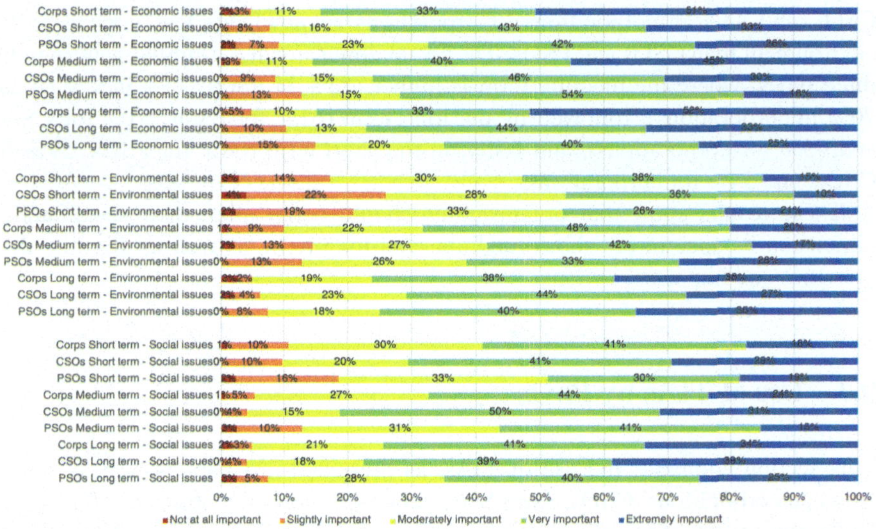

Fig. 9.2 Sustainability priorities of the three types of organisations ($n = 52$ for CSOs, $n = 215$ for corporations, and $n = 45$ for PSOs)

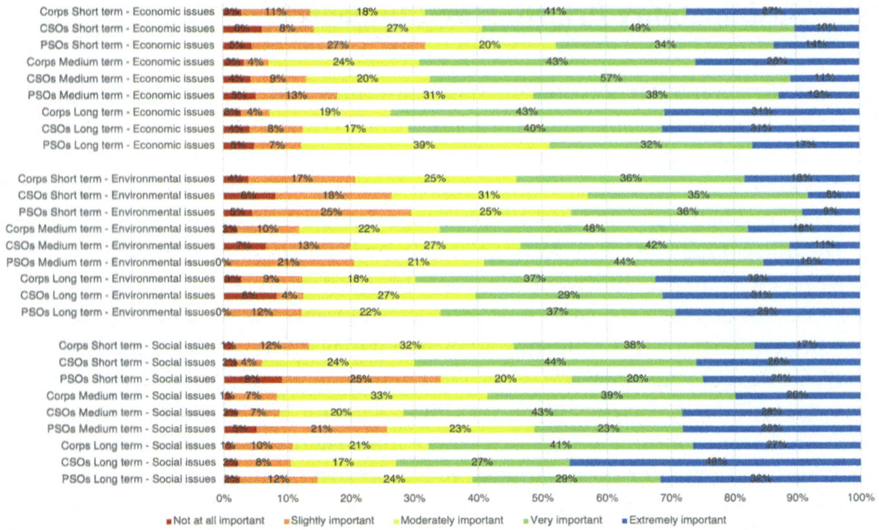

Fig. 9.3 Sustainability impacts of the three types of organisations (n = 52 for CSOs, n = 215 for corporations, and n = 45 for PSOs)

69%, 71% vs 67% in the medium term, and 49% vs 51% in the long term). Corporations have a higher impact in the short term (68%) than in the medium term (59%) or long term (48%). In the environmental dimension, the three types show a similar pattern of higher impacts in the short term (70% in PSOs, 66% in CSOs, and 54% in corporations) and then on the long term (66% in PSOs, 59% in CSOs, and 45% in corporations), being lowest for the medium term ((60% in PSOs, 53% in CSOs, and 43% in corporations). Corporations acknowledge less impacts to the environment than CSOs and PSOs. The social dimension has a similar pattern in the three types, where the medium term is higher (73% in PSOs, 72% in CSOs, and 70% in corporations), followed by the long term (61% for PSOs, 51% for CSOs, and 45% for corporations), and lastly, in the short term (68% for PSOs, 59% for CSOs, and 54% for corporations).

These results show that organisations should better align their sustainability dimension balance in the short, medium, and long term, as well as aligning their priorities and impacts.

9.3 Use of TIAs by PSOs

This section compares the use of TIAs, as presented in Chapters 3, 5, 6, and 7, between corporations, CSOs, and PSOs. Table 9.1 shows the number of TIAs used by the three types of organisations (for an explanation of TIAs, refer to Chap. 3). As can be seen from Table 9.1, the three types of organisations tend to use more

Table 9.1 Number of TIAs used by the three types of organisations

	Corporations	CSOs	PSOs
Too many initiatives (more than 15), i.e. waste of efforts	46.53%	38.00%	40.91%
High number of initiatives used (more than 7), consider combining or removing some	44.55%	42.00%	38.64%
Too few used (less than 3)	2.97%	10.00%	4.55%
Apparent good use of initiatives (between 4 and 6)	5.94%	10.00%	15.91%

Table 9.2 TIA ranking quintile comparison

	Corporations	CSOs	PSOs
Corporate Social Responsibility	1	1	1
Corporate sustainability	1	2	2
Cleaner Production	2	2	2
Eco-efficiency	2	2	2
Circular Economy	3	3	3
Corporate Citizenship	3	3	3
Design for the Environment/eco-design	3	3	3
Environmental Management Systems (ISO 14000 series)	3	4	2
Integrated Management Systems	3	3	3
Life Cycle Assessment	3	3	3
Socially/Sustainable Responsible Investment	3	3	3
Sustainability reporting (GRI report)	2	4	3
Ecolabels	3	4	3
Sustainable supply chain	3	4	3
Environmental Management Systems (EMAS)	4	4	3
Green/sustainable marketing	3	4	4
Green/sustainable chemistry	4	4	4
Industrial ecology	4	4	4
Triple Bottom Line	4	4	4
Sustainability reporting (AA1000)	4	5	4
Factor X	5	5	5
Sustainability reporting (ISO 26000)	5	5	5
Sustainability reporting (SA8000)	5	5	5
The Natural Step	5	5	5

initiatives than would be ideal (around 90% for corporations and around 80% for CSOs and PSOs). PSOs have the highest percentage in the apparent good use of initiatives (between 4 and 6), followed by CSOs, and lastly by corporations.

Table 9.2 shows a comparison of TIA quintile rankings of the three organisation types analysed. It can be seen that the use of TIAs is mostly in the same rank in all three types of organisation, with some exceptions:

- Differences in all types: Sustainability reporting (GRI report) (2 in corporations, 3 in PSOs, and 4 in CSOs) and EMS (ISO 14000) (2 in PSOs, 3 in corporations, and 4 in CSOs).
- One TIA is higher in one type of organisation: Green/sustainable marketing is higher in corporations than in CSOs and PSOs, EMS (EMAS) and higher in PSOs than in corporations and CSOs, and corporate sustainability in higher in corporations than in CSOs and PSOs.
- One TIA is lower in one type of organisation: Sustainability reporting (AA1000) is lower in CSOs than in corporations and PSOs, sustainable supply chain is lower in CSOs than in corporations and PSOs, and eco-labels are lower in CSOs than in corporations and PSOs.

CSOs show a different pattern in TIA usage ranking, with 1 TIA in the highest rank, 3 in the high, 6 in the middle, 9 in the low, and 5 in the lowest. Corporations have 2 TIAs in the highest rank, 3 in the high, 10 in the middle, 5 in the low, and 4 in the lowest. PSOs have 1 TIA in the highest rank, 4 in the high, 10 in the middle, 5 in the low, and 4 in the lowest.

The ratio analyses of TIAs resulted in two to four groups, with corporations having four (low, medium, high, and highest), CSOs with three groups (low, medium, and high), and PSOs with two groups (low and medium). This may indicate that the use of TIAs in corporations is more mature. The TIA cluster analyses resulted in five groups for corporations, four for CSOs, and five for PSOs. The ratio and cluster analyses provide guidance for each type of organisation on how to choose the best combination of TIAs to address the sustainability dimensions.

9.4 Sustainability Change Management

This section compares the sustainability change management efforts, as discussed in Chapters 4, 5, 6, and 7, between corporations, CSOs, and PSOS (Table 9.3). An analysis of these three types of organisations showed that, when ranked, sustainability has been driven, as follows: equally, by a combination of external stimuli and internal factors, mainly by internal factors but with some external stimuli, or mainly by external stimuli but with some internal factors. There are differences in the ranking for these types of organisation (as shown in Table 9.2); whereas corporations are more balanced and internal factors predominate, CSOs are skewed slightly towards internal factors and PSOs to external factors. Solely internal or external factors are rare.

The three types of organisations analysed showed that, when ranked, sustainability has been resisted (see Table 9.4) by a combination of mainly internal factors but with some external stimuli and equally by balanced external stimuli and internal factors. For corporations, resistance was due, mainly, to external stimuli but with some internal factors and is higher ranked than solely by internal factors. The

Table 9.3 Ranking of sustainability driving forces in the three types of organisations

	Corporations	CSOs	PSOs
Solely by internal factors	4	4	
Mainly by internal factors, but with some external stimuli	2	1	3
Equally by external stimuli and internal factors	1	2	1
Mainly by external stimuli, but with some internal factors	3	3	2
Solely by external factors	5	5	

Table 9.4 Ranking of sustainability resistance forces in the three types of organisations

	Corporations	CSOs	PSOs
Solely by internal factors	4	3	2
Mainly by internal factors, but with some external stimuli	1	1	1
Equally by external stimuli and internal factors	2	2	2
Mainly by external stimuli, but with some internal factors	3	4	3
Solely by external factors	5	5	4

opposite appears in CSOs and PSOs. Solely by external factors was ranked the lowest in the three types of organisations.

From Table 9.4, it will be seen that the sustainability driving forces are more balanced between internal factors and external stimuli, whereas the resistance forces tend to be skewed towards internal factors.

Table 9.5 shows the comparison of the sustainability drivers between the three organisation types analysed. It can be seen that the most highly ranked drivers are proactive leadership and moral and ethical obligations (internal drivers), reputation (connecting driver), and increased levels of social awareness (external driver), whereas, the lowest ranked drivers are social crises, institutional shareholders, market demands for non-financial information, NGO activism, shareholder activism, negative publicity, and peer pressure (all external drivers). The remaining drivers are ranked in different positions, depending on the organisation type. The pattern of driver type strengthens the discussion on sustainability driving forces and provides more details per organisation type.

Table 9.6 shows the ranking for starting sustainability incorporation, i.e. the initial changes in the system elements per organisation type. Thus, it can be seen that CSOs present a very different pattern from corporations and PSOs. In CSOs, changes tend to start in service provision and organisational systems, whereas in corporations and PSOs, changes tend to start in management and strategy and governance. Assessment and reporting and operations and production are more important in corporations than in PSOs, where organisational systems and research and development are more important.

Table 9.5 Ranking of the sustainability drivers in the three types of organisations

	Driver type	Corporations	CSOs	PSOs
Proactive leadership	Internal	1	1	1
Increased levels of social awareness	External	2	1	1
Reputation	Connecting	1	1	2
Moral and ethical obligations	Internal	2	1	2
Organisation culture	Internal	1	2	3
Regulation and legislation	External	2	3	1
Collaboration with external parties	External	3	3	1
National and regional policies	External	4	3	1
Avoiding risk	Internal	2	3	4
Customer demands and expectations	External	2	4	3
Environmental crises	Connecting	4	3	2
Economic considerations	Internal	3	3	4
Market positioning	Connecting	2	4	4
Precautionary principle	Internal	2	5	3
Sustainability reports	Connecting	2	5	3
Business case	Internal	3	3	5
Increased students' awareness	External	5	2	4
Market opportunities	Connecting	2	4	5
Sustainability champions	Internal	3	3	5
Access to resources	Connecting	4	4	5
Demands from employees	Internal	5	3	5
Social crises	Connecting	5	4	4
Institutional shareholders	External	5	5	4
Market demands for non-financial information	External	4	5	5
NGOs activism	External	5	5	4
Shareholder activism	External	5	4	5
Negative publicity	External	5	5	5
Peer-pressure	External	5	5	5

Table 9.6 Ranking of sustainability incorporation in the system elements per organisation type

	Corporations	CSOs	PSOs
Management and strategy	1	6	1
Governance	2	6	2
Assessment and reporting	3	6	6
Operations and production	4	6	5
Organisational systems	5	2	3
Supply chains	6	5	6
Research and development	7	3	4
Collaboration	8	4	7
Service provision	9	1	7

Table 9.7 shows the ranking of sustainability institutionalisation (i.e. where the changes have taken place) in the system elements per organisation type. Governance and operations and production are two system elements where changes take place in almost equal (high) ranking in the three types of organisations. Besides these two elements, CSOs tend to have a different pattern of institutionalisation from corporations and PSOs. In CSOs changes tend to take place more in service provision, whereas for corporations and PSOs, changes tend to occur more in management and strategy.

Table 9.7 Ranking of sustainability institutionalisation in the system elements per organisation type

	Corporations	CSOs	PSOs
Management and strategy	1	8	2
Governance	2	2	1
Assessment and reporting	7	5	5
Operations and production	3	3	3
Organisational systems	4	8	6
Supply chains	5	7	4
Research and development	8	6	8
Collaboration	6	4	7
Service provision	8	1	8

Table 9.8 Difference in ranking of sustainability incorporation and institutionalisation in the system elements per organisation type

	Corporations	CSOs	PSOs
Management and strategy	0	-2	-1
Governance	0	4	1
Assessment and reporting	-4	1	1
Operations and production	1	3	2
Organisational systems	1	-6	-3
Supply chains	1	-2	2
Research and development	-1	-3	-4
Collaboration	2	0	0
Service provision	1	0	-1

Table 9.8 shows the differences in ranking between sustainability incorporation and institutionalisation, where it can be seen that the highest changes were for assessment and reporting in corporations (down 4 places), organisational systems in CSOs and PSOs (down 6 and 4 places, respectively), research and development in CSOs and PSOs (down 3 and 4 places, respectively), governance in CSOs (up 4 places), and operations and production (up 3 places in CSOs and 2 in PSOs). In general, the system elements do not change much within organisation types, but they differ between organisation types. It can thus be posited that it is not so crucial where sustainability efforts begin to be incorporated and how they are institutionalised but that all the system elements must, ultimately, be addressed.

9.5 Sustainability Collaboration

This section compares the sustainability collaboration, as presented in Chapters 1, 5, 6, and 7, between corporations, CSOs, and PSOs. Table 9.9 shows the ranking of sustainability collaboration elements per organisation type. As it can be seen, the patterns are quite similar, with Business models and Products in the top rank and Innovation and Services in the bottom rank. PSOs have two elements, Community and Education, which are one rank higher than corporations and CSOs, and corporation has one element, Research, which is one rank lower than CSOs and PSOs.

Table 9.10 shows the ranking of sustainability collaboration benefits per organisation type. It can be seen that two benefits are ranked equally for all organisation types (Decreased time needed to accomplish objectives and Reducing or removing conflicts) and one in the middle (More efficient processes). The other benefits have differences in ranking for the organisation types, with access markets and knowledge, enrich creativity, and transdisciplinary learning tending towards the bottom rank.

Table 9.9 Ranking of sustainability collaboration elements per organisation type

	Corporations	CSOs	PSOs
Business models	1	1	1
Products	1	1	1
Community	2	2	1
Recovery (reduce, reuse, recycle)	2	2	2
Technology	2	2	2
Research	3	2	2
Education	3	3	2
Innovation	3	3	3
Services	3	3	3

Table 9.10 Ranking of sustainability collaboration benefits per organisation type

	Corporations	CSOs	PSOs
Decreased time needed to accomplish objectives	1	1	1
Reducing or removing conflicts	1	1	1
Ability to optimise financial and human capital	2	1	1
Being action orientated	2	1	2
More efficient processes	2	2	2
Offering benefits to all the players	2	1	3
Access markets and knowledge	3	2	3
Enrich creativity	3	3	2
Trans-disciplinary learning	2	3	3

Table 9.11 Ranking of sustainability collaboration challenges per organisation type

	Corporations	CSOs	PSOs
Free-riding	1	1	1
Structural conflicts	1	1	1
Bargaining problems (how to split gains)	1	2	1
Data conflicts	1	1	2
Relationship conflicts	1	1	2
Value conflicts	2	2	1
Risk of jeopardising important and unique resources	2	2	2
Conflicting needs	2	2	3
Coordination problems	2	2	3
Information problems (who benefits)	3	2	3
Operational dependencies between the activities of the different actors	3	3	3

Table 9.11 shows the ranking of sustainability collaboration challenges per organisation type, where it can be seen that three challenges are ranked in the same position in all organisation types, two on the top rank (Free-riding and Structural conflicts), and one in the bottom rank (Operational dependencies between the activities of the different actors). Corporations and CSOs have a similar pattern, with the exception of Bargaining problems which is ranked on top in corporations and in the middle for CSOs, and Information problems, which is ranked in the middle in CSOs and in the bottom in corporations. PSOs differ in six challenges from corporations and CSOs.

9.6 Conclusions

Organisations are an integral part of modern societies, and not, as is oft times considered, stand apart islands. They have a duty to contribute to sustainability, as governments do through the SDGs. There has been considerable research on organisations' contributions to sustainability, most of it on corporations and HEIs. However, such efforts have been mainly on technocentric and managerial ploys and have remained compartmentalised in each organisation type, which has not allowed for comparisons or cross-fertilisation between organisations.

The results of the studies reported in this book show that there are a number of similarities between organisation types but some considerable differences. However, regardless of their type, organisations should recognise that embedding and contributing to sustainability have to encompass the four dimensions of sustainability (economic, environmental, social, and time); use the appropriate combination of TIAs, depending on the organisation type and context; engage the system elements through change management processes, recognising external stimuli, internal factors, drivers for change, incorporation, and institutionalisation; and foster collaboration.

The findings of this book are aimed at providing a complementary perspective on technocentric and managerial approaches, so that organisations can better balance their sustainability priorities and align them with their impacts, use the TIAs more efficiently, engage with change processes, and tap into the considerable benefits of entering into collaboration with other organisations.

The book also presents the theoretical foundation, and the empirical data, for sustainability priorities and impacts, the TIAs, change management, and collaboration to showcase the efforts of CSOs, corporations, and PSOs have undertaken to contribute to sustainability.

Organisations bear a responsibility to make societies more sustainable, which entails introducing changes to their systems, priorities, impacts, and perspectives. This book provides solutions to this through an integrative, holistic perspective on how organisations contribute to sustainability through analyses of sustainability priorities and impacts, TIAs, change management, and collaboration.

References

Corsi, K., & Arru, B. (2020). Role and implementation of sustainability management control tools: Critical aspects in the Italian context. *Accounting, Auditing and Accountability Journal, 34*(9), 29–56. https://doi.org/10.1108/AAAJ-02-2019-3887

Danter, K. J., Griest, D. L., Mullins, G. W., & Norland, E. (2000). Organizational change as a component of ecosystem management. *Society and Natural Resources, 13*(6), 537–547. https://doi.org/10.1080/08941920050114592

Dlouhá, J., Huisingh, D., & Barton, A. (2013). Learning networks in higher education: Universities in search of making effective regional impacts. *Journal of Cleaner Production, 49*, 5–10. https://doi.org/10.1016/j.jclepro.2013.01.034

Domingues, A. R., Lozano, R., Ceulemans, K., & Ramos, T. B. (2017). Sustainability reporting in public sector organisations: Exploring the relation between the reporting process and organisational change management for sustainability. *Journal of Environmental Management, 192*, 292–301. https://doi.org/10.1016/j.jenvman.2017.01.074

Doppelt, B. (2003). Leading change toward sustainability. In *A change-management guide for business, government and civil society*. Greenleaf Publishing.

Dumay, J., Guthrie, J., & Farneti, F. (2010). Gri sustainability reporting guidelines for public and third sector organizations. *Public Management Review, 12*, 37–41.

Dyllick, T., & Hockerts, K. (2002). Beyond the business case for corporate sustainability. *Business Strategy and the Environment, 11*, 130–141.

Gebauer, H., & Worch, H. (2015). Absorptive Capacity (of Organizations). In *International Encyclopedia of the Social & Behavioral Sciences* (Vol. 1, 2nd ed., p. Elsevier). https://doi.org/10.1016/B978-0-08-097086-8.73106-0

Guthrie, J., & Farneti, F. (2008). GRI sustainability reporting by Australian public sector organizations. *Public Money & Management, 28*(6), 361–366.

Henriques, A., & Richardson, J. (2005). *The triple bottom line. Does it all add up?* Earthscan.

Holliday, C. O. J., Schmidheiny, S., & Watts, P. (2002). *Walking the talk. The Business Case for Sustainable Development*. Greenleaf Publishing.

Hörisch, J., Ortas, E., Schaltegger, S., & Álvarez, I. (2015). Environmental effects of sustainability management tools: An empirical analysis of large companies. *Ecological Economics, 120*, 241–249. https://doi.org/10.1016/j.ecolecon.2015.11.002

Jennings, P. D., & Zandbergen, P. A. (1995). Ecologically sustainable organizations: An institutional approach. *Academy of Management Review, 20*(4), 1015–1052. https://doi.org/10.5465/AMR.1995.9512280034

Jones, G. R. (2013). *Organizational theory, design, and change (seventh)*. Pearson Education Limited.

Langer, M. E., & Schön, A. (2003). *Enhancing corporate sustainability. A framework based evaluation tools for sustainable development*. Forschungsschwerpunkt Nachhaltigkeit und Umweltmanagement, Wirtschaftsuniversität Wien.

Linnenluecke, M. K., Russell, S. V., Griffi, A., Linnenluecke, M. K., & Griffiths, A. (2009). Subcultures and sustainability practices : The impact on understanding corporate sustainability. *Business Strategy and the Environment, 18*(December 2007), 432–452. https://doi.org/10.1002/bse

Lodhia, S., Jacobs, K., & Park, Y. J. (2012). Driving public sector environmental reporting. *Public Management Review, 14*(5), 631–647.

Lozano, R. (2008). Developing collaborative and sustainable organisations. *Journal of Cleaner Production, 16*(4), 499–509. https://doi.org/10.1016/j.jclepro.2007.01.002

Lozano, R. (2012). Orchestrating Organisational changes for corporate sustainability. *Greener Management International, 57*, 43–64. https://doi.org/10.9774/GLEAF.3062.2007.sp.00005

Lozano, R. (2018). Proposing a definition and a framework of organisational sustainability: A review of efforts and a survey of approaches to change. *Sustainability, 10*(4), 1157. https://doi.org/10.3390/su10041157

Lozano, R., Ceulemans, K., Alonso-Almeida, M., Huisingh, D., Lozano, F. J., Waas, T., Lambrechts, W., Lukman, R., & Hugé, J. (2014). A review of commitment and implementation of sustainable development in higher education: Results from a worldwide survey. *Journal of Cleaner Production, 108*. https://doi.org/10.1016/j.jclepro.2014.09.048

Pfeffer, J. (2010). Building sustainable organizations: The human factor. *Academy of Management Perspectives, 24*, 34–45. https://doi.org/10.5465/AMP.2010.50304415

Porter, L. W., Lawler, E. E., & I., & Hackman, J. R. (1975). *Behavior in organizations*. McGraw-Hill.

Rogers, E. M. (1962). *Diffusion of innovations (Fourth)*. The Free Press of Glencoe.

Schaltegger, S., & Burrit, R. L. (2005). Corporate sustainability. In H. Folmer & T. Tietenberg (Eds.), *The international yearbook of environmental and resource economics 2005/2006* (pp. 185–222). Edward Elgar.

Scott, W. R., & Davis, G. F. (2015). Organization: Overview. In *International Encyclopedia of the Social & Behavioral Sciences* (Vol. 16, 2nd ed.). Elsevier. https://doi.org/10.1016/B978-0-08-097086-8.73112-6

Soyka, P. A. (2012). *Creating a sustainable organization*. FT Press.

Stacey, R. D. (1993). *Strategic management and organisational dynamics*. Pitman Publishing.

Tschopp, D. J. (2005). Corporate social responsibility: A comparison between the United States and the European Union. *Corporate Social Responsibility and Environmental Management, 12*, 55–59.

Index

A
AA1000, 58, 61, 92, 94, 108, 118, 120, 132
A Blueprint for Survival, 8
Anthropocentric, 11, 25
Assessment and reporting, 2, 22, 30, 31, 39, 54, 58–62, 80, 97, 110, 113, 123, 125, 128, 133, 135

B
Barriers to change, 32, 78, 80, 81, 83
Benefits to stakeholders typology, 23, 24
Biodiversity, 8, 58
Brundtland Report, 10, 13
Business models for sustainability, 53, 54

C
Carbon dioxide, 6, 7
Change management, 2, 3, 21, 30, 62, 79–85, 100, 109, 113, 120, 125, 132, 137, 138
Change pathways, 76, 79
Change processes, 1–3, 31, 32, 75, 128, 138
Chemical Leasing (ChL), 45, 53
Circular economy (CE), 30, 39–41, 92, 94, 108, 118, 120
Civil society organisations (CSOs), 1–3, 19, 32, 84, 89–100, 112, 125, 127, 129–138
Cleaner production (CP), 39, 41, 92, 94, 106, 108, 118, 120
Closed systems, 19
Collaboration, 1–3, 28, 29, 57, 77, 83, 97–100, 110–113, 122–125, 136–138
Collaboration benefits and challenges, 30
Communication, 27, 28, 51, 57, 59, 78, 79
Companies, 127
Company responsibility choice typology, 23
Competition, 28, 42, 57, 75, 78
Congruence, 22, 28
Connecting drivers, 82
Conventional economists' approach, 13
Co-operation, 28
Co-ordination, 28
Corporate and Industrial Voluntary Initiatives for Sustainability (CIVIS) framework, 62, 63
Corporate citizenship (CC), 48, 92, 94, 106, 108, 118, 120
Corporate social responsibility (CSR), 48–51, 92, 94, 106, 108, 118, 120
Corporate sustainability (CS), 23, 24, 30, 51, 82, 92–94, 106, 118, 120
Corporate sustainability strategic influence typology, 24
Corporations, 1–3, 12, 19, 23, 32, 48–50, 55, 57, 58, 84, 99, 103–113, 125, 129–138
Curricula, 89

D
Design for the environment (DfE), 39, 42–44, 92, 94, 108, 118
Developmental change, 76
Drivers for change, 2, 78–79, 81, 82, 85
Drivers for sustainability, 80

E
Earth Summit, 10
Eco-centric, 11

R. Lozano, *Toward Sustainable Organisations*, Strategies for Sustainability, https://doi.org/10.1007/978-3-030-99676-5

Eco-efficiency, 39, 42–46, 92, 94, 106, 108, 118, 120
Ecolabelling, 55–56
Ecolabels, 94, 108, 118, 120, 132
Economic dimension, 39, 90–92, 104, 105, 112, 117, 129, 137
Education institutions, 19, 32, 89, 127, 128
Electricity, 6
Enronian type, 25
Environmental and social accounting (ESA), 58
Environmental dimension, 13, 39, 53, 90, 91, 104, 105, 117, 129, 130, 137
Environmental management systems (EMS), 58–59, 61, 94, 106, 108, 118, 120, 132
EU EcoManagement and Audit Scheme (EMAS), 59, 61, 94, 108, 118, 120, 132
External drivers, 81, 82

F
Factor X, 39, 44, 94, 106, 108, 109, 118, 120
First Tier Sustainability Equilibrium (FTSE), 13
Fordian type, 25

G
GDP per capita, 6
Global Reporting Initiative (GRI) report, 62, 94, 108, 118, 120, 132
Global Reporting Initiative (GRI) Sustainability Reporting Guidelines, 62, 115
Governance, 21, 30, 31, 51, 52, 54, 57, 62, 84, 97, 100, 110, 113, 115, 122, 125, 128, 133–135
Green chemistry (GC), 39, 45–46
Green indulgences, 25
Green marketing, 56
Green Revolution, 5
Green/sustainable chemistry, 94, 108, 118, 120
Greenhouse gas emissions, 7, 42, 89
Groups, 19, 20, 22, 23, 25–27, 31, 51, 56, 62, 77, 78, 128, 132

H
Hans Carl von Carlowitz, 7
Higher education institutions (HEIs), 2, 32, 84, 89, 99, 116, 137
Holistic approach, 13, 62
Holistic perspective, 3, 51, 80, 81, 138
Holistic type, 25
Human resources, 53

I
Incorporation, 3, 30, 79, 83, 84, 97, 100, 110, 111, 113, 122, 123, 125, 128, 133–137
Individuals, 19, 20, 22, 26–28, 30, 31, 56, 77, 78, 103, 128
Industrial ecology, 39, 46–47, 94, 108, 118, 120
Institutional framework, 21, 81
Institutionalisation, 3, 79, 81, 82, 84, 97, 100, 110, 111, 113, 122, 123, 125, 134–137
Insular type, 25
Integrated management systems (IMS), 58, 60, 94, 108, 118, 120
Integrational approach, 13
Inter-generational approach, 13
Inter-group interactions, 26
Internal drivers, 82
International Organisation for Standardization (ISO), 59
Inter-organisational interactions, 26
Inter-personal interactions, 26
Investors, 12, 23, 52, 83
ISO 14000, 59, 61, 94, 106, 108, 118, 120, 132
ISO 26000, 61, 94, 108, 109, 118, 120

L
Leadership, 21, 49, 57, 77, 78, 82, 89, 95, 109, 122, 133
Leverage, 20, 28, 77, 78
Life cycle assessment (LCA), 39, 47–48, 59, 94, 108, 118, 120
Life cycle management (LCM), 47–48
Limits to Growth, 8
Lower management, 21

M
Management and strategy, 2, 21, 31, 39, 48–55, 84, 97, 110, 113, 122, 125, 128, 133, 134
Managerial approaches, 1, 3, 137
Managerial efforts, 100, 113, 125
Managerial ploys, 2, 32, 62, 79, 128, 137
Middle management, 20–21
Millennium Development Goals (MDGs), 10
More sustainability-orientated state (MSOS), 81
Multi-dimensional Sustainability Influence Change memework, 22

N
9Rs, 40
Non-environmental degradation approach, 13
Non-governmental organisations (NGOs), 19, 24, 44, 49, 50, 57, 82, 83, 89, 90, 97, 110, 122

O
Open systems, 19
Operations and production, 2, 21, 30, 31, 39–48, 51, 53, 54, 84, 97, 100, 110, 113, 122, 125, 128, 133–135
Orchestrating Change for Corporate Sustainability model, 81
Organisational change management, 2, 30, 80, 85
Organisational change management for sustainability, 2
Organisational changes, 2, 30, 75, 77–85
Organisational culture, 30
Organisational focus change, 76
Organisational sustainability, 1, 2, 30, 32, 80, 127, 128
Organisational sustainability collaboration framework, 29
Organisational sustainability framework, 31, 128
Organisational systems, 21, 30–31, 51, 54, 97, 100, 110, 113, 122, 123, 125, 133, 135
Our Common Future, 10

P
Philanthropic type, 25
Philanthropy, 48, 50, 56
Planned changes, 77
Planned organisational change, 77
Population, 6, 8, 9
Population bomb, 8
Procurement and marketing, 51
Production, 5
Production and consumption, 6, 40
Public sector organisations (PSOs), 1–3, 19, 32, 84, 99, 112, 115–125, 127–138

Q
Quality of life, 6, 14, 49

R
Rank-and-file positions, 21
Rate of change, 2, 31, 75, 85, 128
Registration, Evaluation, Authorisation & Restriction of Chemicals (REACH), 46
Religious organisations, 2, 89, 90, 99
Research and development, 21, 28, 97, 100, 110, 123, 133, 135
Research on sustainability, 89
Resistance forces to sustainability, 95
Resistance to change, 2, 78–79, 81–83, 85

S
SA8000, 58, 61, 92, 94, 106, 108, 109, 118, 120
Semi-closed systems, 20
Semi-open systems, 20
Service provision, 21, 31, 97, 100, 110, 123, 133, 134
Shareholders, 24, 25, 49–51, 83, 97, 110, 122, 133
Silent Spring, 8
Social dimension, 91, 104, 117, 129, 130, 137
Social marketing, 56
Stakeholder emphasis typology, 24
Stakeholder engagement, 48, 49
Stakeholder focus, 76
Stakeholders, 1, 2, 11, 12, 19, 23–26, 31, 32, 40, 48–54, 56, 60, 61, 76, 85, 127, 128
Stakeholder scope typology, 24
Strategy and management, 30
Strong sustainability, 11
Supply chains, 2, 21, 31, 32, 39, 53–57, 79, 94, 97, 108, 110, 113, 118, 120, 123, 125, 128, 132
Sustainability assessment and reporting, 80–81
Sustainability change drivers, 81–83
Sustainability change management, 95, 132
Sustainability change resistance, 83
Sustainability changes, 85, 97, 110, 111, 122, 123
Sustainability collaboration, 29, 98, 111, 112, 123, 124, 136
Sustainability drivers, 96
Sustainability impacts, 91, 105, 117, 129, 138
Sustainability priorities, 1–3, 129, 138
Sustainability priorities and impacts, 2, 3, 113, 125, 129, 130, 138
Sustainability reporting (SR), 58, 60–62, 106, 115, 116, 118, 120
Sustainable business models (SBMs), 48, 52–54
Sustainable development (SD), 1, 5, 7, 9, 10, 13, 14, 30, 32, 54, 57, 60, 89

Sustainable Development Goals (SDGs), 10, 137
Sustainable livelihoods (SL), 55–57
Sustainable marketing (SM), 55, 56, 94, 108, 118, 120, 132
Sustainable organisation, 32
Sustainable society, 9, 10
Sustainable supply chains (SSC), 55, 57, 106, 108
Sustainable/socially responsible investment (SRI), 48, 52, 94, 108, 120
System elements, 1–3, 20–23, 30, 32, 48, 58, 62, 75, 79, 84, 85, 97, 110, 111, 122, 123, 128, 133–137
Systems thinking, 20, 83

T
Tangible resources, 53
Technocentric approaches, 1–3, 11, 32, 62, 79, 137
Technocentric efforts, 100, 113, 125, 128
The natural step (TNS), 48, 54, 92, 94, 106, 108, 109, 118, 120
Three ‘pillars, 13
3Rs, 40
Time dimension, 2, 13, 30, 32, 51, 53, 75, 85, 127, 137
Tools, Initiatives, and Approaches (TIAs), 1–3, 32, 39–63, 92–95, 100, 106–108, 113, 118–121, 125, 130–132, 137, 138
Top management, 20, 57, 78
Transitional change, 76
Transitional period, 81
Triple bottom line (TBL), 13, 48, 54–55, 94, 108, 118, 120
Two-tiered sustainability equilibria (TTSE), 13

U
UN Conference on Human Environment, 8
United Nations Conference on Environment and Development (UNCED), 10
Upper management level initiatives, 81

W
Weak sustainability, 11
World Council of Churches, 9, 90
World Summit on Sustainable Development, 10, 45

The manufacturer's authorised representative in the EU is Springer Nature Customer Service Centre GmbH, Europaplatz 3, 69115 Heidelberg, Germany. If you have any concerns regarding our products, please contact ProductSafety@springernature.com

Printed and bound by CPI Group (UK) Ltd, Croydon, CR0 4YY
01/10/2026
02225883-0010